THE COURAGE

TO FACE COVID-19

THE COURAGE TO FACE COVID-19

Preventing Hospitalization and Death While Battling the Bio-Pharmaceutical Complex

John Leake

Peter A. McCullough, MD, MPH

Foreword by Robert F. Kennedy Jr.

Skyhorse Publishing

Skyhorse Publishing books may be purchased in bulk at special discounts for sales promotion, corporate gifts, fund-raising, or educational purposes. Special editions can also be created to specifications. For details, contact the Special Sales Department, Skyhorse Publishing, 307 West 36th Street, 11th Floor, New York, NY 10018 or info@skyhorsepublishing.com.

Visit our website at www.skyhorsepublishing.com.

10 9 8 7 6 5 4

Library of Congress Cataloging-in-Publication Data is available on file.

Hardcover ISBN: 978-1-5107-7680-7

Cover design by TheBigLogic.com

Printed in the United States of America

This book is dedicated to all men and women who have shown the courage to face COVID-19 by choosing critical reasoning over fear, principle over expediency, and factual reality over propaganda. For the last two years they have paid a heavy price for obeying their conscience. Men and women like them have always struggled to lead mankind out of the dark. They are the backbones of all free nations.

Very few have been granted the ability to prove of great and lasting service to mankind, and with few exceptions, the world has crucified and burned its benefactors. Still, I hope you will not tire in the honorable struggle that remains for you.

—Dr. Louise Kugelmann to Dr. Ignaz Semmelweiss

Contents

Foreword

by Robert F. Kennedy Jr.

My old friend Dr. Peter McCullough is an internist and cardiologist who is the most published physician in his specialization in history. He is probably the most knowledgeable person alive on the ambulatory management of COVID-19. He published in August and December of 2020 seminal articles that were the most downloaded reports during the entire pandemic on the outpatient sequential use of various drugs including ivermectin, hydroxychloroquine and monoclonal antibodies in the treatment of COVID-19. He has been harshly punished and silenced by the medical cartel but has nevertheless continued to speak out. He is one of the great, great heroes of the last two years. John Leake is a historian and author who has written award-winning books. The two of these gentlemen have collaborated to write this nonfiction chronicle of the deliberate, purposeful, and lethal suppression of early treatment.

One of the most sinister aspects of this story is the convergence of power centers simultaneously moving to block public access to life-saving drugs and into a 200-billion-dollar vaccine syndicate. It was all tied in with government controls, the intelligence apparatus, and the media. Dr. Peter McCullough is the most dangerous person facing the Bio-Pharmaceutical Complex. His coming forward has really changed the calculus of this global crisis.

Preface

Soon after SARS-CoV-2 started spreading in the United States, I sensed the official pandemic response was full of fraudulent misrepresentations. Public health officials and the corporate media falsely depicted the virus as a grave threat to the entire population, including the young and healthy. The infectious agent was portrayed as an insidious and unassailable monster of which everyone should be equally terrified. Nothing (we were told) in the entire pharmacopoeia would work against it, and we were discouraged from even trying anything. Safe old drugs like hydroxychloroquine and ivermectin that had been prescribed to millions for other diseases were suddenly characterized as "dangerous" and made inaccessible to COVID-19 patients.

This created a big problem for people who were at significant risk of severe COVID-19—namely, older people, especially those over 65, and those with underlying medical conditions such as diabetes. I suspected that a crime was being committed against this large segment of the population—a crime with elements of fraud and negligent homicide. As the reader will see, public health agencies started maligning and restricting access to certain repurposed, generic drugs right as evidence was emerging that these medications help to prevent hospitalization and death from COVID-19. To me, this raised some disturbing questions that became the basis of my inquiry. If you deny sick people medicines that could prevent them from sickening further to the point of hospitalization and death, are you not at least partly to blame if they indeed wind up going to hospital and dying? Is withholding medicine from a sick man any different from withholding a life ring from a man who has fallen overboard in high seas?

Initially I had no idea why our public officials and media would suppress repurposed, generic drugs, and I was astounded that anyone would want to do such a perfidious thing. Nevertheless, the totality of circumstances indicated that this crime was being committed, and I resolved to investigate it. I got my first lead on or around May 12, 2020, when I saw Dallas County Judge Clay Jenkins giving a COVID-19 press briefing. He was standing in front of a color-coded graph titled "Today's COVID-19 Risk Level." On the far left was code red on which was printed "Stay Home Stay Safe," which had been the risk level all spring. On the far right was code green on which was printed "New Normal Until Vaccine." In other words, only with the arrival of a COVID-19 vaccine would we be able to enjoy a semblance of the old normal we'd always known. This indicated that, for our public health officials, it was a forgone conclusion that a safe and effective vaccine was coming, and that it was the only solution.

But how could they possibly know this? The first human trial of Moderna's new vaccine had only begun on March 16. What if it and other vaccines in development didn't prove to be safe and effective? To me, this sounded an awful lot like a done deal, regardless of how the trials went. It reminded me of an old TV ad from the eighties in which a purveyor of custom mufflers contrasts his product with that of a shop that only sells mufflers in one standard size. At the competitor's garage, an alarmed customer asks, "But what if that muffler doesn't fit my car?" To this query, the shop owner points to a group of chimps pounding the muffler with baseball bats. "We'll make it fit," he says.

Yet another feature of the official narrative that made me queasy was the frequent proclamation that public health officials were "following the science"—as though "science" were a fixed entity in their possession. I knew from forty years of studying history and medical history that every generation overestimates its understanding of nature. Proper scientific inquiry has always given us glimpses into how much we don't know. As Oliver Wendell Holmes Sr. put it, "Science is the topography of ignorance. From a few elevated points we triangulate vast spaces, inclosing infinite unknown details." What did our public health officials know about the novel virus? They spoke a great deal about contagion control and computer modeling of its hypothetical spread, but they never mentioned what treating doctors in the field were observing.

I sensed that if anyone was going to lead us out of the disaster, it would have to be a talented and diligent doctor. Likewise, I figured I could only pursue my investigation so far without the assistance of a medical authority. Ideally, he or she would combine extensive clinical practice with academic medical distinction. Where would I find such a person? And if I found him, would he have the time and inclination to work with me? Shortly after Halloween, someone sent me a YouTube video of Dr. Peter McCullough, standing in a park, wearing a jogging outfit. In the video he was talking about the early treatment of COVID-19 in order to prevent hospitalization and death. This struck me as both marvelous and odd. For months, our public health officials had told us there is no treatment for COVID-19. And yet, here was the vice chief of Internal Medicine at Baylor University Medical Center telling us the disease is treatable. His message gave me hope—the first time I'd felt it since March. Another notable thing about the video was his statement that he was "in Glencoe Park," which happens to be about a mile from my home.

I followed Dr. McCullough and watched recordings of his US Senate testimony on November 19, 2020, and his Texas State Senate testimony on March 10, 2021. Then I watched a recording of his *Tucker Carlson Today* interview on May 7, 2021. That is when I decided to contact him and invite him to my own studio interview. He spoke so eloquently, with such encyclopedic knowledge, that his interview required no editing. The astonished video director recommended releasing it uncut for the world to see.

This conversation was the first of many. I got to know him not only as a compassionate doctor (who frequently took calls from sick patients in the evening and made house calls), but also as a devoted family man and loyal friend. Beyond his boundless passion for medical scholarship, he is deeply interested in the entire human condition and the integrity of our Constitutional Republic.

One fine evening in July 2021, we met at a Spanish tapas restaurant, and over dinner he suggested that we write a book together. During this same dinner, he told me that I wasn't the only investigative author who'd realized the official policy response to COVID-19 was terribly wrong. Robert F. Kennedy Jr. and Peter R. Breggin, MD, were also writing books about it, and he'd given extensive interviews to both authors. He told me they shared many of my suspicions and had arrived at many of the same conclusions. That all three of us were, independently, seeing many of the same things indicated to him that what we were seeing was real. At the end of dinner, we agreed that, with his experience as a medical doctor and mine as a true crime writer, we could tell this story together.

Because I'm an experienced nonfiction author, I have written our account in my narrative voice. Nevertheless, we worked together on the design and scope of the book. Many extended passages are taken verbatim from his oral accounts. He provided me with much of my source material, introduced me to other key players, offered innumerable suggestions for augmenting and elaborating the story, and edited the typescript. The final product is the result of a unique and fruitful partnership.

Prologue

On October 19, 2019, the Johns Hopkins Center for Public Security, in collaboration with the Bill and Melinda Gates Foundation and the World Economic Forum, conducted a Pandemic Simulation Exercise. As Johns Hopkins described the event:

> The center's latest pandemic simulation, Event 201, dropped participants right in the midst of an uncontrolled coronavirus outbreak that was spreading like wildfire out of South America to wreak worldwide havoc. As fictional newscasters from "GNN" narrated, the immune-resistant virus (nicknamed CAPS) was crippling trade and travel, sending the global economy into freefall. Social media was rampant with rumors and misinformation, governments were collapsing, and citizens were revolting.[1]

A video of the simulation opened with a description of the disease:[2]

> Infected people got a respiratory illness with symptoms ranging from mild, flulike signs to severe pneumonia. The sickest required intensive care; many died. . . . A Pandemic Emergency Board has been convened to respond to the CAPS pandemic.

The simulation video featured a debate on the "GNN" network between immunologist Dr. Yabani Bello and epidemiologist Dr. Rea Blakey:

> **Dr. Yabani Bello**: We have heard about Caps-like viruses in animals and humans for decades, but we have not been successful in developing a licensed vaccine. And sure there are new technologies that are going to help, but it's going to be difficult. . . . Even if we discover a good vaccine candidate, we are

> starting from scratch, and it takes time to test safety and efficacy, typically years.
>
> **Dr. Blakey**: We simply cannot rely on these old timelines and processes. This is a crisis. We have to move beyond these issues. It may be complicated, difficult, but if we dedicate all available resources, it can happen. Also, keep in mind, we need effective treatments sooner rather than later. Extranovir is an effective antiviral drug. . . . We need to start treating people immediately.
>
> **News Host**: Our US affiliate has just released polling results on public expectations for a vaccine. The majority of Americans expect a vaccine to be available in two months, and 65 percent of those polled are eager to take the vaccine, even if it's experimental.

The simulation cut back to the Pandemic Board, with medical countermeasure expert Matt Watson contrasting the time-consuming development of a new vaccine with the readily available, generic, antiviral extranovir, an HIV drug repurposed to treat CAPS:

> As we just heard, there is no existing vaccine that appears effective against the newly discovered CAPS virus. Governments, companies, and scientists around the world are working intensively in order to develop one, but it's highly unlikely that a vaccine could be developed, tested, manufactured, and distributed in less than a year, and it's likely to take much longer. . . . The antiviral drug extranovir has demonstrated some efficacy against the CAPS virus. It reduces the severity of illness and saves lives in those affected, but it does not prevent infection like a vaccine would.

Dr. Watson did not mention any concern about the drug's safety for use against CAPS. That one million HIV patients took it indicated it was safe.

Later in the discussion, Dr. Timothy Evans drew his colleagues' attention to a new organization called CEPI that was founded in Davos, Switzerland, in 2017 by the Bill & Melinda Gates Foundation, the World Economic Forum, and the Wellcome Trust. As Dr. Evans told the Pandemic Board:

> CEPI, which was established three years ago to accelerate vaccine development exactly in these settings where a coronavirus is, in fact, one of its indications and first areas of focus. That may not be common knowledge.

Event 201 and other planning exercises since 2017 were uncannily prescient. Many participants said they knew a devastating pandemic was coming. It was only a matter of time, and CEPI—the Coalition for Epidemic Preparedness Innovations—was prepared for it. Indeed, CEPI was already developing vaccines for a virulent coronavirus outbreak.

CHAPTER 1

The Plague Is Coming

February 23, 2020

I looked at my Facebook newsfeed and saw a post from an Italian friend. The Italian government had just issued Decree Law Number 6, directing state authorities throughout the country to take extraordinary measures to contain SARS-CoV-2. The novel respiratory virus had spread rapidly in the Greater Milan area, and the government was trying to stop it or at least slow it down. With horror I read reports of people falling ill with severe acute respiratory syndrome (SARS) and being rushed to hospital, many of them subsequently dying. The unfolding drama in Italy caught my attention because I'd lived in the country and loved its culture and people. Since 2000 my home base had been Vienna, Austria, but over the years I'd rented apartments in Venice, Florence, and Rome. Being a freelance author enabled me to enjoy this mobility.

The first twenty years of the 21st century had been an exceptionally peaceful and prosperous time in European history, and I'd had the good fortune to live in Europe during this period. As I read about the Italian government imposing measures that would soon be referred to as "lockdown," I feared this happy era was ending and might never resume. *Lockdown,* as in a prison lockdown when the inmates have gotten out of hand, seemed antithetical to all Europe's progress since 1945. How could this be happening in 2020?

My formal education is in history and philosophy, but I've long been interested in the history of medicine. My great-great-grandfather, William

Reid Wilson, was a surgeon at several of the Civil War's bloodiest battles. After the war he served as the Dallas city health officer and undertook pioneering initiatives for controlling yellow fever and smallpox. At the end of his tenure, he submitted a comprehensive public health plan to the Dallas city council for building a community hospital and community clean water and sewage systems. His entire plan was implemented.

His wartime diary got passed down to my grandmother, and I read it when I was a boy. His description of a field hospital "resembling one of the lowest circles in Dante's *Inferno*" with a multitude of horribly wounded soldiers shrieking in agony in a miasma of blood and putrefaction made a vivid impression on me, and it made me curious about the general state of medicine at the time. I wanted to learn more about how wounded soldiers were anesthetized before surgery, and how surgeons like my great-great-grandfather sterilized his instruments. So began my lifelong fascination with how discoveries in medicine were made.

An obvious lesson from medical history is that, at any given time, our knowledge of the human body and its pathology is limited. Understanding has constantly evolved through a multiplicity of observations, experiments, and debates. Error has always been a big part of the process. Sometimes errors were detected and corrected quickly; sometimes they persisted for a while. Successive generations have often looked back at the errors of the past and wondered how their medical forefathers could have been so quaint. Future generations will doubtless think the same about us.

My work as a nonfiction (true crime) author had led me deep into the literature of forensic medicine. Forensic pathologists and scientists played major roles in my first two books. I'd also done some translation work (from German to English) for a doctor at the Vienna Institute of Forensic Medicine. During the 19th century, its faculty made key contributions to the development of modern medicine. While visiting I sometimes thought of Professor Jakob Kolletschka, the institute director who died in 1847 after he was accidentally cut by a student's scalpel during a postmortem examination. Kolletschka's colleague and friend, Professor Ignaz Semmelweis, hypothesized that ptomaine (German *Leichengift* or "cadaver poison") was transferred from the cadaver to Professor Kolletschka's body by the scalpel. He further postulated that students were transferring "cadaver particles" from their hands (*an der Hand klebende Cadavertheile*) to pregnant women whom they examined in a maternity clinic following their anatomy class. Wondering if this could be the cause of high rates of puerperal fever (an illness that emerges right after childbirth) in the clinic, he instructed the students to wash their hands with chlorinated lime before they examined the women.

Soon puerperal fever mortality in the clinic dropped precipitously, thereby opening a seminal chapter in the development of the germ theory of disease. Unfortunately for Dr. Semmelweis—and for many patients who could have benefitted from his insight sooner rather than later—his theory was vehemently rejected by many of his era's eminent medical authorities.[3] Branded a quack and abandoned by everyone, including his wife, he ended up dying alone in a Vienna insane asylum, probably from the effects of an injury he sustained from an abusive staff member.

The moral of the Semmelweis story is clear: orthodoxy in medicine is deadly. The eminent authorities at one time may be revealed to be catastrophically wrong a few years later. The Semmelweis story is also a dreadful example of Groupthink—the belief in something not because it has been subjected to rigorous analysis, but because one's group embraces it. Groupthink may be especially potent when it's propagated and reverberated by perceived authorities on the subject. Because scientific knowledge is always evolving, there can be no final arbiter of scientific truth.

Reports of the contagion in Italy reminded me of literature I'd read about the great Spanish Flu Pandemic of 1918, which had struck Vienna and other crowded European cities very hard. The famous modern painter, Egon Schiele, and his pregnant wife, Edith, died of it. She was twenty-five; he was twenty-eight. While himself gravely ill, he sketched a poignant portrait of her on her deathbed, gazing at him. She died on October 28, 1918; he followed her to the grave three days later.

Understanding of the Spanish Flu was retarded by the opinion of the German bacteriologist Dr. Richard Pfeiffer. Twenty years earlier he'd claimed that influenza was caused by a bacterium that he isolated from the noses of influenza patients in 1892—a bacterium that he named *Bacillus influenzae*.[4] Pfeiffer, a close colleague of Robert Koch, made many important discoveries in bacteriology and immunology, so few scientists in 1918 were inclined to question his authority, even when they found no evidence that his bacillus was the causative agent. In 1931, the American virologist Richard Shope conducted experiments that led him conclude that a virus caused the Spanish Flu (though the severe pneumonia was probably caused by a secondary bacterial infection).[5]

Just over a century after the Spanish Flu struck Europe, COVID-19 hit Milan. Nowhere in any of the reporting did I see a single reference to *treating* the novel infection. It was a terrifying contagion, and it seemed the only policy response was to try to contain it. The fear of it seemed to preclude any consideration of how to treat it. Had medical science really made no progress in treating respiratory viruses since influenza A was isolated in 1933?

I did a bit of research and learned that this novel virus, called SARS-CoV-2, was closely related to SARS-CoV-1, which had, in 2003, also emanated from China and spread to other countries.[6] I still remembered CNN's frequent coverage of the outbreak, which occurred at the same time as the US invasion of Iraq in March of 2003. For about four months, SARS-CoV-1 had posed a major threat. Again, this raised the question: Had medical science made *zero* advances in treating the disease? Surely, I thought, this can't be true.

A quick review of the literature on antiviral therapies showed that, since the 1980s, advances had been made in developing compounds that inhibit viruses such as HIV, HSV, and HPV. Antiviral medications for influenza viruses had also been developed. These included oseltamivir phosphate, inhaled zanamivir, intravenous peramivir, and oral baloxavir marboxil. All of these were thought to be most effective when taken *as early as possible* after exposure to the flu.

A Google search for "Treatment for SARS-CoV-1" yielded seven papers by distinguished authors, published in major medical journals between 2003 and 2014, about the antiviral properties of chloroquine, hydroxychloroquine, and zinc. The lead author of a 2005 paper was a top scientist at the CDC.[7] I read these papers and wondered why no one was talking about them. It was as though an entire body of research on a promising therapy for a deadly viral infection had been forgotten.

Reporting from Italy revealed that SARS-CoV-2 differed from the 1918 Spanish Flu in an extremely favorable way—namely, it wasn't killing children and young people. Whereas the average age of death from the Spanish Flu was 28,[8] initial reports from Italy indicated that SARS-CoV-2 was mostly killing people who had already reached or exceeded life expectancy.[9] That it was sparing young people struck me as an enormous blessing, for there is nothing more devastating than the death of children. Basic logic suggested that if the virus was primarily lethal for people over seventy, the most sensible policy was to protect this cohort, especially those living in nursing homes, which are notoriously vulnerable to respiratory diseases.

In mid-March, my girlfriend had to evacuate her mother from a top-tier nursing home in Dallas because COVID was running through the facility. This happened at a time when only about two hundred cases had been detected in Dallas County. In other words, the one policy response that seemed to make sense—i.e., taking extraordinary measures to prevent viral transmission to the elderly—didn't happen. It was my first clue that our public health authorities either didn't know what they were doing or were powerless to stop the virus from spreading no matter what they did.

At this same time in March, I realized that I wouldn't be returning to Vienna anytime soon. I'd flown home to my native Dallas for Christmas and was still visiting my mother in January when COVID arrived in Europe. Now it was arriving in the United States, and it looked like we would soon be going the way of Italy—that is, into lockdown. On March 13, President Trump declared a nationwide emergency. On March 15, state authorities began shutting down businesses and schools and ordering their residents to stay at home.

One hundred and twenty-eight years after the Russian botanist Dimitri Ivanovsky discovered viruses, it seemed that mankind was still helpless against them. All we could do, according to our Dallas County Judge Clay Jenkins, was "Shelter in Place"—an expression that seemed to emphasize how little humans had advanced since Paleolithic times, when we sheltered in caves from nature's hostile elements. To me the phrase conjured images of T-Rexes roaming our streets, or bands of armed marauders going from house to house. Only this mortal threat was far more insidious because it was invisible. Everyone, it seemed, should be equally afraid of it.

Was "sheltering in place" really the best we could do? Anthony Fauci, who was availing himself of something like Papal Infallibility, seemed to think so. Even though young people weren't at risk of severe illness, they were still ordered to stay home. Prohibited from going to school, socializing, and working, they saw their lives put on hold. I often thought of the irony that, over the course of the 20th century, the US government had sent hundreds of thousands of college-aged men to their deaths in wars abroad to fight dictatorships in Europe and Asia. Now the current generation of youth was living under a public health dictatorship in the United States.

Young people were not at risk from the virus, but they could still contract it and pose a risk to their parents if they brought it home. We were told that everyone, even if they seemed healthy, could have the virus. No one could trust that anyone else was virus-free. Thus, the rationale for confining the young was to protect the older generations. Gregarious youngsters were viewed with suspicion, and if they were caught trying to steal time together in a park or other open space, they were often reprimanded by their frightened elders.

I found it troubling that an entire generation of young people was being constantly told *to be afraid*—both for themselves and for adults whom they could infect. It seemed to me that, while children and teenagers should be taught to develop a good sense of prudence, it was a moral catastrophe to condition them to be afraid all the time, for by far the most important virtue—the foundation of all others—is courage.

"Courage, mankind's finest possession," as the Spartan poet Tyrtaeus called it, is what enables us to do the right thing, to face danger, and to enjoy life. People who live in fear are miserable. As Shakespeare put it in the mouth of Julius Caesar, "Cowards die many times before their deaths; The valiant never taste of death but once." How, I wondered, would the constant conditioning to be afraid affect the development of our young people?

Listening to our state officials invoke "Emergency Authority" (the oldest trick in the dictator's book), I feared for the health of our Constitutional Republic. Throughout history, politicians have relished threats to public safety, as these have enabled them to assume the mantle of "Great Leader." The danger of foreign invaders is the easiest to comprehend and mobilize against. James Madison, the author of the US Constitution, warned of the age-old tendency of politicians to invoke foreign dangers when intractable problems at home cause social unrest.

In the spring of 2020, I often thought about Madison as I watched press briefings by New York Governor Andrew Cuomo and Los Angeles Mayor Eric Garcetti. Neither, it seemed to me, knew what he was talking about, but both clearly enjoyed being in the limelight and wielding enhanced power. To be sure, they frequently referred to "public health officials" who were guiding their dictates, enabling them to "follow the science." This suggested that true executive power resided in unelected bureaucrats.

On May 15, Mayor Garcetti announced that LA residents would be allowed to go to the city's beaches that coming weekend, but only on the wet sand area. Tanning and hanging out on the dry sand would not be allowed.

"But the wet sand area—if you need to get in there to swim, to surf . . . that is something I hope we can earn again," he proclaimed.[10]

Mayor Garcetti's statement reminded me of the newly minted dictator in the Woody Allen film *Bananas*, who at his inauguration pronounces that "all citizens will be required to change their underwear every half hour. Underwear will be worn on the outside so we can check."[11]

I wondered if, somewhere out there in our vast medical complex, there was an eminent doctor who took a different view of the matter—someone who had the knowledge, imagination, and courage to fight the virus instead of ordering us to hide from it. As I would discover in the weeks ahead, there were several. The man who would ultimately assume a leadership role lived about a mile from my family home.

CHAPTER 2

Preparing for Battle

Dr. Peter McCullough was, at the beginning of 2020, the vice chief of Internal Medicine and program director of Cardiology at the Baylor University Medical Center in Dallas. In January a Chinese friend sent him a video, shot in a crowded, panicked hospital waiting room in Wuhan. Amongst a throng of patients, one individual collapsed. This, it seemed, was what can happen with SARS (severe acute respiratory syndrome). One minute the patient is conscious and breathing; the next he decompensates and falls to the floor. *Holy smokes, that's scary*, McCullough thought. The sender indicated the Chinese government didn't want such videos getting out, which is why he sent it through a virtual private network.

In January and February, McCullough monitored the unfolding drama—first in Wuhan, and then in Milan—and tried to learn everything he could about the virus. He figured it was no coincidence that both cities had high urban density, with people living in close quarters in high-rise apartment buildings with poor ventilation. If a family member brought home the infection, it was likely to spread to everyone in the dwelling, who were reinoculated by the enclosed air containing the virus.

Thinking the SARS-CoV-1 outbreak in 2003 might offer a clue, he called a friend in Ottawa, Canada, Dr. Manish Sood, whose hospital had admitted infected patients. Dr. Sood wasn't encouraging. All staff could do was try to protect themselves. For ninety days he'd worn an N-95 mask and hoped for the best. As for their patients—most of those in his hospital had sickened to the point of requiring mechanical ventilation, on which many had died. None of the Canadian hospitals had developed a treatment protocol. Interest in doing so seemed to wane after the outbreak ran its course. Later McCullough learned that SARS-CoV-1 had generated massive interest in the virus and developing a vaccine for it, but relatively little interest in developing a treatment for it.

Seventeen years later SARS-CoV-2 hit the United States. It was a terrible irony of medical history that no one knew what to do about it. During the first 150 years of modern medical practice, doctors and scientists chiefly concerned themselves with understanding and treating infectious diseases. However, by the year 2020, few doctors practicing in the developed world had experienced an infectious disease outbreak. Advances in public sanitation, clean drinking water, antibiotics, and vaccines had eliminated the contagions that had plagued Europe and America for centuries. In 1963, the year after McCullough was born, the trivalent oral poliovirus vaccine was commercialized and finished off polio in the United States. Consequently, there probably wasn't a doctor or nurse at Baylor who'd contended with the high risk of being infected from a patient with a fatal pathogen.

A conspicuous exception were nurses at Presbyterian Hospital, a few miles north of Baylor, who'd treated a patient infected with Ebola virus in 2014. Between September 26 and November 5, the world watched Dallas as it scrambled to deal with a potential Ebola outbreak, the first recorded in US history. The patient, who'd flown from Liberia to Dallas, ultimately succumbed to the disease. An investigation revealed that hospital personnel had made grave errors in handling the patient that had resulted in two nurses being infected (both of whom survived) and that could have exposed numerous others.[12] As soon as the CDC was alerted, it flew officials to Dallas to guide the response. Total disaster was averted, and the incident provided Dallas County Judge Clay Jenkins—the county's top emergency management authority—with a lot of cherished press.

Six years later at Baylor University Medical Center, discussions of how to handle the coming storm focused on Personal Protective Equipment (PPE) for staff. Early in the pandemic, before the first COVID-19 patient arrived, there was increasing talk of masks. Though it wasn't clear how effective they were at preventing transmission, they were invested with a growing symbolic and psychological significance. How this came about included a dramatic tipping point that illustrated how collective sentiment can suddenly shift 180 degrees. Before deep fear really set in, only two doctors in cardiology started wearing masks outside of the operating room and catheterization lab. One day, they approached the nursing station while still wearing their masks. Some nurses found it jarring and reported the doctors to hospital administration. The chief medical officer then issued an e-mail directing staff not to wear masks outside of designated areas. A few weeks later, when masks became all the rage, this e-mail became a running joke about the flip-flopping of health authorities across the United States.

McCullough would never forget a staff meeting in which a young and handsome doctor, originally from South America, volunteered to serve on the front line of the expected wave of COVID patients. The data from China indicated that infected young people fared far better than people over fifty.

"A disaster is coming," he said, "a mass casualty event, and some of us will have to be in harm's way to deal with it. I and my group are still relatively young, so we are happy to volunteer." Everyone present was deeply moved, and it reminded McCullough of young soldiers volunteering for a dangerous mission. Other young doctors and nurses were inspired by his example, and they too volunteered to serve on the front line. McCullough felt very proud of them, and proud to be part of their institution.

In mid-March, a 56-year-old man flew from New York to Dallas and felt feverish when he got home. In the days that followed he fell increasingly ill and ultimately landed in the Baylor ICU with acute respiratory distress, where he was put on a ventilator and died two days later. He was Baylor's first COVID-19 patient, and his precipitous decline and death had a chilling effect on the staff. Though the ventilator failed to save him, the staff didn't know what else could have been done for him. The ineffectiveness of mechanical ventilation implied that the virus must have caused some pathology that was different from typical severe pneumonia.

The experience with this first patient was consistent with reports about SARS-CoV-1 in 2003, when patients experienced sudden acute respiratory distress. Even more frightening was a 22-year-old nurse who contracted the infection from an older patient. She sickened to the point of requiring mechanical ventilation in prolonged ECMO (Extra Corporeal Membrane Oxygenation), an advanced form of life support in which blood is routed outside of the body and oxygenated independently of the lungs and mechanical ventilation. Though cases like hers were relatively rare, they were terrifying to behold. As a result, they prompted hospitals nationwide to put patients on ventilators soon after they were admitted. People who were conscious and talking when they arrived were nevertheless sedated, paralyzed, intubated, and placed in isolated rooms on ventilators. Their family members couldn't see them, and even doctors didn't visit them, only nurses. Given how many of these patients ultimately died, the procedure struck

McCullough as a dark dead-end street. *Surely*, he thought, *there's some way to treat this thing before it requires mechanical ventilation.*

To prepare for the predicted onslaught, Baylor halted most of its diagnostic and elective procedures. McCullough's department—the Baylor Scott & White Heart and Vascular Hospital—closed its heart catheterization lab. His patients (most of whom needed cardiovascular care) were informed that, barring an emergency, consultations were to be conducted by phone or video conferencing. In-person interaction between doctors and patients, doctors and nurses, and doctors with each other was greatly reduced. Virtually overnight, much of medical practice shifted to phones and computer screens.

With his patient workload reduced, McCullough had more time to do what he'd always loved and was exceptionally good at—medical research. A practicing board-certified internist and cardiologist, he was also a professor of Medicine at Texas A&M University, president of the Cardiorenal Society of America, and editor in chief or senior associate editor of three major academic journals, and on the editorial board of more than a dozen journals. A marathon runner who abstained from alcohol, he could work and study from 5:00 a.m. to midnight for days on end. By the year 2020, he had published over 600 peer-reviewed academic medical papers and over a thousand medical communications. In addition to his MD, he'd been awarded eight additional medical certifications from numerous societies.

His research specialization was cardiovascular medicine with a focus on the interaction between the heart and kidneys. Two key components of his work were epidemiology and pharmacology—that is, understanding the prevalence and risk factors for diseases, and understanding how to evaluate the efficacy and safety of drugs for treating them. Over the years he had organized and supervised several clinical trials, and he had chaired or served on data safety monitoring boards for over two dozen randomized trials.

In mid-March, his chief of cardiology, Dr. Kevin Wheelan, posed a big question.

"What are we going to do to fight this monster? If our nurses get wiped out, we're done for." McCullough had already been thinking about the bigger problem of how to prevent hospitals nationwide from being overrun. Research teams in China were reporting favorable results from treating COVID patients with hydroxychloroquine—a drug that had been FDA-approved since 1955 for malaria prophylaxis, as well as for the treatment of lupus and rheumatoid arthritis. In India, the state medical councils recommended that frontline medical workers take the drug as prophylaxis.

"The only thing I'm seeing in the literature is the antimalarial hydroxychloroquine," McCullough replied. "I think we should take the Indian medical council's recommendation and give it a try."

And so, he and Dr. Wheelan decided to conduct a clinical trial of hydroxychloroquine as prophylaxis against COVID-19 for their hospital workers. McCullough applied to Baylor for a research grant and wrote a protocol to apply to the FDA for an investigational drug application for a new use for hydroxychloroquine. This was a major undertaking that required writing a sixty-page document under extreme time pressure and preparing his staff, which in turn talked with FDA officers for most of a weekend in order to get his application number assigned.

Because of the extraordinary circumstances, there was no time to produce a placebo. It was also necessary to ask staff to decide who wanted to take the drug and who didn't, which meant it wasn't randomized. Nevertheless, with about 200 in the experimental group and 200 in the control, the trial seemed likely to yield meaningful results. Generally, it was the older doctors and nurses who volunteered to take the prophylaxis. The younger ones tended to prefer taking their chances without it. Weekly nasopharyngeal swab PCR tests were obtained from both groups to determine the presence or absence of the virus.

At the time, this endeavor seemed the most natural course of action. The trial could protect the hospital's workers and provide valuable data (and regular testing for staff, which was, early in the pandemic, difficult to obtain). Little did McCullough know he was embarking on a journey that would ultimately lead him to becoming a central figure in the greatest conflict since the Cold War—a battle in which he would be pitted against an array of powerful interests and institutions, including his own. As he would soon learn, this battle had already begun in France, where a kindred spirit was squaring off against the French national medical establishment.

CHAPTER 3

Gandalf of Marseille

Professor Didier Raoult is sometimes referred to in the French press as "Gandalf of Marseille" because of his resemblance to the famous wizard in the Lord of the Rings trilogy. The 68-year-old's appearance—tall, powerfully built, with shoulder-length hair and white mustache and goatee—expresses his larger-than-life character and biography. Others have opined that he looks like a medieval knight in a white lab coat. Still others have likened him to Panoramix the Druid in the *Asterix* comic series about a village of Gauls who resist Julius Caesar's Roman occupation. The latter comparison is apt, because one of Professor Raoult's most conspicuous traits is his indomitable spirit of independence.

Born in 1952, Didier Raoult spent his early years in Dakar, French West Africa (now Senegal), where his father, a military doctor with the rank of colonel, was posted. Malaria was endemic to the region, so he and his family took hydroxychloroquine as a prophylaxis. In 1961 his family moved back to Marseille. He briefly attended high school in Nice and Briançon and then dropped out to join the French Merchant Marine, where he worked on a ship called the Renaissance. Again, returning to Marseilles, he studied literature for a year and then switched to medicine. He earned his MD and then a PhD in biology.

It's tempting to think he was destined to be an infectious disease researcher, achieving international renown reminiscent of Louis Pasteur's. His father, Dr. André Raoult, was head of a commission in French West Africa to investigate the main causes of death in the territory. His maternal great-great-grandfather, Louis Paul Le Gendre, was also a doctor and infectious disease researcher who published 260 academic papers and chaired the French Society for the History of Medicine.[13] Following in the footsteps of his distinguished ancestors, Dr. Raoult made the research and treatment of infectious diseases his life's calling. A May 12, 2020, feature in the *New York Times* introduced him to American readers as follows:

> Raoult . . . has made a great career assailing orthodoxy, in both word and practice. "There's nothing I like more than blowing up a theory that's been so nicely established," he once said. He has a reputation for bluster but also for a certain creativity. He looks where no one else cares to, with methods no one else is using, and finds things.[14]

A tireless researcher, he has published 2,300 papers and is the most cited microbiologist in Europe. He and his team have discovered 468 species of bacteria—about 1/5 of all those named and described. The bacteria genus Raoultella was named in his honor. He is probably best known as the discoverer of the so-called giant virus, so large it had previously been mistaken for an intracellular bacterium. He has won 13 major awards and is a commander of the National Order of Merit.

In 2010 he was instrumental in founding a group of university research hospitals (IHUs)—including his own in Marseille, called the Institut Hospitalo-Universitaire Méditerranée Infection. These IHUs were established as foundations with access to private donations and independence from the state. Though the French state has frequently honored him for his achievements (he is a member of the Legion of Honor), in recent years he'd clashed with heads of national medical institutions.

Especially tense has been his relationship with Dr. Yves Lévy, who served as director of the National Institute of Health and Research from 2014 to 2018. An immunologist, Dr. Lévy spent his formative years researching HIV, seeking an elusive vaccine against the virus, and working in various research units of the National Institute of Health. Dr. Lévy was also part of the French delegation that inaugurated the BSL-4 Lab at the Wuhan Institute of Virology in 2017. The lab was conceived in 2003 during the SARS-CoV-1 outbreak and constructed by the French biotech company bioMerieux SA pursuant to a cooperative agreement between France and China.[15] The CEO of bioMerieux from 2007 to 2011 was Stéphane Bancel. In 2011 it seemed like an extraordinary—perhaps even quixotic—decision to leave this plum position to become the CEO of the small, Cambridge, Massachusetts, start-up, Moderna. At the time, the company had one employee and was exclusively focused on developing mRNA therapeutics.

"When I resigned from my last company, bioMerieux, to start on this journey at Moderna, I told my wife there was only a 5 percent chance it would work out," Bancel stated in a December 2020 interview.[16]

In many respects, Dr. Yves Lévy's career paralleled that of Dr. Anthony Fauci, director of the National Institute of Allergy and Infectious Diseases (NIAID) in the United States, who also devoted many of his formative years to HIV research and vaccine development. In 2016, Drs. Lévy and Fauci were appointed members of the UN's Global Health Crisis Task Force. According to Professor Raoult, who stated his opinion to the French press, "the pursuit of a vaccine for HIV is a fantasy that has cost billions and will never be realized."[17]

This remark and others about the National Institute of Health—spoken with Professor Raoult's characteristic lack of diplomacy—did not endear him to Dr. Lévy. Their long-simmering "reciprocal detestation" erupted into open conflict in 2017, when Director Lévy's wife, Agnès Buzyn, became the French minister of health. Shortly thereafter she attempted to change the foundation status of IHUs to that of public-interest groups under the executive direction of the National Institute of Health. Professor Raoult cried foul and accused Minister Buzyn of pursing policy in the interest of her husband, and not in the interest of medical science.

"The IHUs are an issue of authority and territory for Yves Lévy. He would like to lead them from Paris," Professor Raoul declared to the French weekly newsmagazine, *Marianne*. "Yves Lévy gave orders to everyone; he believes he can make us obey. Great scientists do not obey anyone."[18] In the end, Professor Raoult prevailed, and IHUs retained their original legal status. Professor Lévy's tenure at the National Institute of Health concluded in 2018, but his wife remained the minister of health, where she was serving in 2020 when COVID-19 arrived in France.

As a treating physician and researcher, Professor Raoult had long been interested in drug repurposing—that is, experimenting with existing medicines, already approved for treating some diseases, to determine if they are useful for treating others. As the *New York Times* profile put it:

> Hundreds and hundreds of molecules have already been approved for human use by the Food and Drug Administration. Hidden among these, Raoult contends, are various unanticipated cures. "You test everything. . . . You stop

> pondering; you just look and see if, by chance, something works. And what you find by chance, it'll knock you on your derrière."

For decades he'd been interested in the antimicrobial properties of hydroxychloroquine. Though the drug was originally developed to prevent and treat malaria, it had proven to have an array of a salutary effects. In the 1990s, he gave hydroxychloroquine a try against Q fever, an often-deadly disease caused by an intracellular bacterium, which, like viruses, multiplies within the host's cells. Raoult discovered that hydroxychloroquine reduces acidity inside the cells, thereby inhibiting bacterial growth. He commenced treating Q fever with a combination of hydroxychloroquine and doxycycline. Later he used the same combination to treat Whipple's disease, another frequently fatal condition, also caused by an intracellular bacterium. Professor Raoult's combination therapy is now widely regarded as the standard treatment for both diseases.

In 2003 he followed the outbreak of SARS-CoV-1 in China and was familiar with the literature on the antiviral properties of chloroquine and hydroxychloroquine against the novel virus. In 2007 he reviewed the evidence in his own paper and concluded that hydroxychloroquine might be "an interesting weapon to face present and future infectious diseases worldwide." In 2018, he reported that azithromycin had potent activity in cells infected with Zika virus.

When SARS-CoV-2 broke out in 2020, again emanating from China, he wanted to know how Chinese doctors were treating the disease. In February, he saw reports from multiple Chinese research teams stating that chloroquine and hydroxychloroquine were reducing disease severity and accelerating clearance of the virus. The People's Hospital of Wuhan University posted a notice on its website that none of the 178 patients admitted so far had lupus, suggesting the hydroxychloroquine they were taking to treat lupus may have prophylaxis value against COVID-19. On February 19, a pharmacology research team in Qingdao, China, published a report in English that on February 17, the State Council of China held a news briefing in which it announced that chloroquine phosphate had demonstrated marked efficacy and acceptable safety in treating COVID-19. After positive in vitro studies, several clinical trials were conducted to test chloroquine and hydroxychloroquine for the treatment of COVID-19.

> Thus far, results from more than 100 patients have demonstrated that chloroquine phosphate is superior to the control treatment in inhibiting the

> exacerbation of pneumonia, improving lung imaging findings, promoting a virus-negative conversion, and shortening the disease course, according to the news briefing. . . . Given these findings . . . the drug is recommended for inclusion in the next version of the Guidelines . . . issued by the National Health Commission of the People's Republic of China.[19]

For Professor Raoult, this report was cause for jubilation. Facing a pandemic that could inflict untold suffering and death on mankind, it appeared that a well-known, safe, cheap, and easy-to-manufacture drug was effective at treating it. Wanting to share this good news, on February 25 he posted a video on YouTube titled *Coronavirus: Game Over!*

"It's excellent news—this is probably the easiest respiratory infection to treat of all," Raoult said.

He then formulated his own treatment protocol. He chose hydroxychloroquine because it is less toxic than its analogue, chloroquine. He combined it with azithromycin because of the antiviral effect he'd observed in using it against the Zika virus, and for its value against secondary bacterial infections. As for the safety profile of these drugs, they'd been around for decades, with billions of doses taken and well tolerated. Both hydroxychloroquine and azithromycin are on the World Health Organization's Model List of Essential Medicines.

At the beginning of March, Professor Raoult announced that he would test and treat anyone who came to his hospital. Come they did, lining up at the entrance. In order to handle the surge, he and his staff erected field tents outside to examine sick patients and give them the combination therapy.

He then briskly performed a clinical study to evaluate the effect of hydroxychloroquine on respiratory viral loads. Presence and absence of the virus at Day 6 post inclusion was the end point. Fourteen were treated with hydroxychloroquine, and six were treated with a combination of hydroxychloroquine and azithromycin. The control group was composed of sixteen untreated patients from another center and cases refusing the protocol. On Day 6 of the trial, fourteen of the sixteen control patients still tested positive for the virus. Only six of the fourteen patients receiving hydroxychloroquine tested positive on Day 6. *All 6 patients treated with a combination of hydroxychloroquine and azithromycin had cleared the virus.*[20]

Professor Raoult understood the results were not proof of the drug's efficacy in treating COVID-19. Nevertheless, the study indicated that the cocktail of hydroxychloroquine + azithromycin substantially reduced viral load in the upper respiratory tract. This provided a strong rationale for using it to treat people while performing additional case studies to examine its efficacy. Professor Raoult quickly published the results of this study in the *International Journal of Antimicrobial Agents.*

The academic medical community's response was conspicuously tepid, perhaps even hostile. Given the apparent direness of the situation facing mankind, one might have expected at least an openness to Professor Raoult's results. The psychiatrist and author Dr. Norman Doidge pointed out that the response was a conspicuous example of "unwishful thinking."[21] Normal people with no unstated agenda could only wish for a cheap generic drug with an excellent safety profile that could save millions of lives. Conversely, those guilty of unwishful thinking dismissed hydroxychloroquine's potential benefit even before it had been examined.

Shortly before Professor Raoult's study was published, he shared it with two independent researchers in the United States: an ophthalmologist named Dr. James Todaro at the Beaumont Hospital in Dearborn, Michigan, and a lawyer named Gregory Rigano. They too had seen the Chinese reports, and they'd also seen a February 13 report in the *Korea Biomedical Review* headlined "Physicians work out treatment guidelines for coronavirus":

> The COVID-19 Central Clinical Task Force, composed of physicians and experts treating the confirmed patients across the nation, held the sixth video conference and agreed on these and other treatment principles for patients with COVID-19. . . . if patients are old or have underlying conditions with serious symptoms, physicians should consider an antiviral treatment. If they decide to use the antiviral therapy, they should start the administration as soon as possible. . . . As chloroquine is not available in Korea, doctors could consider hydroxychloroquine 400mg orally per day, they said.[22]

On March 13, Todaro and Rigano published their own paper on Google Docs in which they presented the results of the South Korean and Chinese studies recommending chloroquine and hydroxychloroquine for treating COVID-19.[23] On March 16, Elon Musk retweeted a reference and link to this paper, thereby rapidly publicizing it.[24]

On March 19, Rigano went on Fox News's *Tucker Carlson Tonight* to talk about Professor Raoult's study of hydroxychloroquine, claiming it showed "a 100 percent cure rate." He concluded by stating that the study was scheduled to appear the next day in *The International Journal of Antimicrobial Agents*, so viewers could see it for themselves.

Because Rigano was not affiliated with a major academic medical institution and didn't work for a federal health agency, he lacked medical authority. Mainstream media pundits proclaimed it was absurd for a lawyer to be presenting possible COVID-19 treatments. And yet, the paper that he coauthored with Dr. Todaro was a perfectly competent presentation of the South Korean and Chinese studies, and when he appeared on Tucker Carlson, it was to convey the results of a study conducted by the most cited microbiologist in Europe. Why were the American mainstream media, apart from Fox News, hostile to Rigano's March 13th paper (coauthored with Dr. Todaro) and his good news about Professor Raoult's study on March 19?

The South Korean, Chinese, and French studies were not final proof of the efficacy of chloroquine and hydroxychloroquine in treating COVID-19, but they offered hope and a start for further research. That they were instantly shot down suggested that the legacy media did not *want* these drugs to be useful in treating the disease.

Instead of deriding Rigano and Todaro for their lack of authority, the legacy media should have asked why US federal agencies *weren't* talking about the South Korean, Chinese, and French studies. Why weren't they, and what treatment possibilities were they talking about?

CHAPTER 4

A Vaccine in Record Speed

On March 16—the same day that Elon Musk retweeted a reference to the Rigano-Todaro paper—the National Institute of Allergy and Infectious Diseases (NIAID) announced that it was beginning a clinical trial of a new investigational vaccine at the Kaiser Permanente Washington Health Research Institute (KPWHRI) in Seattle:

> The vaccine is called mRNA-1273 and was developed by NIAID scientists and their collaborators at the biotechnology company Moderna, Inc., based in Cambridge, Massachusetts. The Coalition for Epidemic Preparedness Innovations (CEPI) supported the manufacturing of the vaccine candidate for the Phase 1 clinical trial.
>
> "Finding a safe and effective vaccine to prevent infection with SARS-CoV-2 is an urgent public health priority," said NIAID Director Anthony S. Fauci, M.D. "This Phase 1 study, launched in record speed, is an important first step toward achieving that goal." The investigational vaccine was developed using a genetic platform called mRNA (messenger RNA). The investigational vaccine directs the body's cells to express a virus protein that it is hoped will elicit a robust immune response. The mRNA-1273 vaccine has shown promise in animal models, and this is the first trial to examine it in humans.
>
> Scientists at NIAID's Vaccine Research Center (VRC) and Moderna were able to quickly develop mRNA-1273 because of prior studies of related coronaviruses that cause severe acute respiratory syndrome (SARS) and Middle East respiratory syndrome (MERS). Coronaviruses are spherical and have spikes protruding from their surface, giving the particles a crown-like appearance.

> The spike binds to human cells, allowing the virus to gain entry. VRC and Moderna scientists already were working on an investigational MERS vaccine targeting the spike, which provided a head start for developing a vaccine candidate to protect against COVID-19. Once the genetic information of SARS-CoV-2 became available, the scientists quickly selected a sequence to express the stabilized spike protein of the virus in the existing mRNA platform.[25]

Record speed indeed. The causal agent of COVID-19 wasn't known until Chinese researchers announced they'd identified it on January 7, 2020. A draft copy of the genome was made available to researchers on January 11, and the first reported US case was on January 21 in Seattle.[26] The WHO did not declare COVID-19 a pandemic until March 11—five days before NIAID and Moderna began their human clinical trial.[27] Clearly this partnership had been developing their new mRNA vaccine technology against coronaviruses such as SARS and MERS for some time. According to Moderna's website, the company had been working on mRNA technology for a decade, investing tens of millions in its development. Now it seemed they had an opportunity to deploy it on a global scale.[28]

NIAID's March 16, 2020, announcement not only reported a promising new vaccine technology to combat COVID-19, but also a declaration from Anthony Fauci—who assumed the countenance of the nation's chief public health advisor—that the vaccine his institute had developed (and made a substantial investment in) was an urgent health priority. As long as no drugs were available to inhibit SARS-CoV-2 and prevent the COVID-19 illness it caused, the new vaccine was apparently mankind's only hope and therefore justified enormous resources for its development. Dr. Fauci did not mention in the announcement that the NIH (a government research funding institution) had not only invested in the vaccine for promoting public health, but also claimed to coown the patent and therefore stood to share the royalties from its commercial exploitation.[29]

During the Event 201 Pandemic Simulation on October 19, 2019, several participants believed that a repurposed antiviral medication was the best hope for immediately addressing the emergency because it would take at least a year to develop a vaccine. However, Dr. Timothy Evans drew his colleagues' attention to the fact that CEPI—the Coalition for Epidemic Preparedness Innovations—was already working on a coronavirus vaccine. Now CEPI "supported the manufacturing of the vaccine candidate for the Phase 1 clinical trial." This was the first step in executing the "Preliminary Business Plan" that CEPI published in November 2016.

CHAPTER 5

"The Opportunity"

CEPI was established under Norwegian law as a not-for-profit association with headquarters in Oslo and offices in London and Washington, DC In its Articles of Association, it describes itself as "an international multi-stakeholder initiative supported by governments, international organisations, industry, public and philanthropic funders, academia and civil society groups."[30] The association was formally launched in Davos, Switzerland, in January 2017 by the Bill & Melinda Gates Foundation, the World Economic Forum—which also founded the Global Alliance for Vaccines and Immunization (GAVI) in 2001—the Wellcome Trust, and the governments of Norway and India. Its primary mission is "to co-ordinate funding and stimulate R&D for vaccines against emerging infectious diseases."[31]

At the heart of this funding coordination enterprise is soliciting enormous donations from governments—i.e., taxpayer money—and channeling the funds to pharmaceutical companies engaged in developing vaccines. In November 2016, CEPI published its "Preliminary Business Plan"—a prospectus that it sent out to donors and participants. Its executive summary set forth **The Challenge** and **The Opportunity.**[32]

> **The Challenge**
> As the recent SARS, MERS, Ebola and Zika outbreaks demonstrate, new diseases can emerge quickly and unexpectedly. . . . To ensure robust and effective private sector participation in future outbreaks, industry will require a reliable risk/reward sharing system, a prioritization system for EIDs, and a clear development pathway for emergency-use vaccines.

> **The Opportunity**
> CEPI . . . will rationalize and accelerate research and development responses to new outbreaks by coordinating resources of industry, governments, philanthropic organizations and NGOs, prioritizing development goals, and facilitating the advanced development of vaccines for EIDs.

The "Preliminary Business Plan" is dedicated entirely to vaccine development. Not once does the sixty-page document refer to treating emerging diseases. Under the section **Cost Coverage**, the business plan states: "Vaccine developers who contribute with dedicated capacities should be reimbursed for their direct and indirect costs." Under the section **Shared benefits**, the business plan states:

> It is anticipated that vaccines developed with CEPI support will not be profitable. In the event that a vaccine developed with CEPI support does develop economic value, agreements between CEPI and the vaccine developer will ensure either that CEPI's investment is reimbursed or that the economic value is shared through royalties or other risk sharing agreements.

In other words, vaccine manufacturers will be fully reimbursed for their costs, and they won't make a profit until one of their vaccines developed with CEPI "does develop economic value," at which point they will share the royalties with CEPI. The association's not-for-profit status meant that any royalties it earned by economically valuable vaccines it supported would be tax free.

The Preliminary Business Plan's last five pages presented CEPI's membership roster—a who's who of prominent persons in academia, government health agencies, pharmaceutical companies, and NGOs. All of them were given "The Opportunity" presented by the next infectious disease outbreak that was sure to come. With the arrival of SARS-CoV-2 three years later, the opportunity was upon them, and the time had come to seize it.

NIAID's March 16, 2020, announcement signaled that Dr. Fauci's institution and CEPI were initiating their business plan of developing the vaccine solution instead of the repurposed antiviral solution. Treating COVID-19 simply wasn't in the business plan. This explains why Gates and Fauci and friends weren't interested in the potential value of hydroxychloroquine—an old generic drug that costs approximately $6.00 for a five-day therapeutic course. This was the equivalent of telling the Pentagon and Lockheed

Martin that their new $1.6 trillion F-35 fighter jet program would not be necessary because a cheap modification of existing jets would achieve the same result.

CHAPTER 6

Unwishful Thinking

Mainstream media reporting suggested that *no one* in America's vast complex of university medical centers and public health agencies had heard about the potential value of hydroxychloroquine until March 19, but this can't be true. Among researchers all over the world, hydroxychloroquine was known as one of the most useful drugs ever formulated. Its natural ancestor, quinine, was discovered by the Quechua people of Peru, who ground the bark of the cinchona tree into powder and mixed it with sweetened water as a tonic against diarrhea and malaria. A Jesuit apothecary in Lima named Agostina Salumbrino (1564–1642) observed the Quechua using it, so he sent a sample to Rome, where malaria was endemic to the swampy areas around the city.[33] This was the beginning of quinine being used for malaria prophylaxis and treatment throughout Europe and its international colonies, saving millions of lives over the centuries.

During World War II, the US military gave chloroquine—a synthetic derivative of quinine with lower toxicity—to millions of US soldiers deployed in regions affected by malaria. In addition to malaria prophylaxis, the drug cleared up rashes and inflammatory arthritis. In 1946, Bayer Labs discovered that by adding a molecule from the hydroxyl group, chloroquine's toxicity could be further reduced while conserving its efficacy.[34]

Building on research conducted by various teams, published in the years 2003–2009, a group of Israeli scientists published a 2012 paper enumerating the drug's array of salutary effects as an antiinflammatory, antithrombotic, antimicrobial (including viral), and immune modulator. Among other positive effects, they noted that it decreased macrophage-mediated cytokine production, especially interleukin IL-1 and IL-6.[35] As several research teams would discover during the COVID Pandemic, high IL-6 levels are strongly associated with severe disease, indicating the therapeutic value of compounds that decrease IL-6 production. Hydroxychloroquine also showed promise in reducing glucose intolerance in diabetic patients. Hydroxychloroquine's potent antiinflammatory property had long been well known to rheumatologists, and as ENT (Ear, Nose, and Throat) and pulmonary doctors had long observed, inflammation is an aggravating factor for respiratory tract diseases.

During the COVID-19 Pandemic, treating physicians discovered that hydroxychloroquine's antiviral property was enhanced by combining it with zinc. This observation was consistent with a 2010 paper by Princeton Professor A.J. te Velthuis and UNC Chapel Hill Professor Ralph Baric (and colleagues) published in PLOS Pathogens in which they "demonstrate that the combination of Zn(2+) and PT [pyrithione] at low concentrations . . . inhibits the replication of SARS-coronavirus (SARS-CoV). . . ."[36]

This was notable for two reasons. First, zinc and pyrithione (a common organosulfur compound) have long been known for their antimicrobial properties. For decades zinc pyrithione has been used to treat dandruff and seborrheic dermatitis. Second, one of the paper's authors, Ralph Baric, would later achieve notoriety for his collaboration with the Wuhan Institute of Virology in conducting gain of function research on bat coronaviruses.[37]

Following the outbreak of SARS-CoV-2 in 2020, Professor Baric received much publicity for his role in the development of mRNA vaccine technology, the experimental drug remdesivir, and for his coauthorship of a favorable study on Merck's antiviral molnupiravir.[38] All of these new drugs were still experimental. In none of Professor Baric's writings or interviews since 2020 has he advocated the zinc ionophore therapy about which he published in 2010.

In 2014, Julie Dyall—a researcher at Dr. Fauci's NIAID—and several distinguished colleagues published a major work titled *Repurposing of Clinically Developed Drugs for Treatment of Middle East Respiratory Syndrome Coronavirus Infection,* in which they concluded that hydroxychloroquine and several other compounds demonstrated activity against MERS-CoV and SARS-CoV.[39] On March 18, 2020, a Chinese research team published a study in *Nature* showing that hydroxychloroquine is effective in inhibiting SARS-CoV-2 infection in vitro.[40] The in vitro studies didn't prove efficacy in humans, but they showed that hydroxychloroquine was the strongest of antiviral candidates.

As already noted, by March 19, 2020, multiple clinical studies and observational reports—in South Korea, China, India, and France—had been published. All showed benefit, with no adverse cardiac events, from using chloroquine or hydroxychloroquine to prevent or treat COVID-19. The drug was therefore recommended by official expert panels in South Korea, China, and India. The silence of the US healthcare system about these facts was deafening.

And so, it was left to President Trump to communicate this promising information. It would be hard to imagine a more unsuitable messenger. Dr. Fauci was in an infinitely better position of medical authority to break the news, but as he'd already declared in his agency's March 16 press release, he was interested in his vaccine solution, and not in the repurposed, antiviral solution. At the March 19 Coronavirus Task Force press briefing, President Trump mentioned chloroquine and hydroxychloroquine as potential game changers in the fight against COVID:

> It is known as a malaria drug and it's been around for a long time, and it's very powerful. But the nice part is, it's been around for a long time, so we know that if things don't go as planned, it's not going to kill anybody. When you go with a brand-new drug, you don't know that that's going to happen. You have to see and you have to go long test. But this has been used in different forms, very powerful drug in different forms, and it's shown encouraging—very, very encouraging—early results, we're going to be able to make that drug available almost immediately.[41]

Though factually accurate, it was a clumsy presentation, and the press arrived at the next day's press briefing ready to attack the president's good news. One reporter asked Dr. Fauci, who was standing next to President Trump, if hydroxychloroquine was indeed a promising drug for treating COVID-19.

"The answer is no," Fauci replied. "The evidence you're talking about . . . is anecdotal evidence. . . . It was not done in a controlled clinical trial. So you really can't make any definitive statement about it."[42]

This was a decidedly awkward moment, characterized by the media as Fauci "throwing cold water on Trump's game changer." The president tried to smooth it over by adding: "We'll see. We're going to know soon." Another reporter then aggressively pressed the president about his claim that the drug had shown promise. His tone was accusatory verging on hostile.

"Is it possible that your impulse to put a positive spin on things may be giving Americans a false sense of hope?"

"No—"

"With this not-yet-approved drug," the reporter interrupted.

"Such a lovely question," Trump replied. "Look, it may work; it may not work. I feel good about it."

The reporter's question was a conspicuous example of unwishful thinking. His greatest concern, it seemed, was the purported risk of giving Americans "false hope," and not whether the cheap generic drug could help them. Either he was already convinced that hydroxychloroquine didn't work, or he didn't want it to work. Neither position made sense.

The next day, undeterred by the "cold water" thrown on his hopeful announcement, Trump tweeted a reference to Professor Raoult's study in the *International Journal of Antimicrobial Agents*. Just as the official medical establishment in Paris did not rejoice at Professor Raoult's announcement of the promising therapy, the official medical establishment in Washington, DC, and the US mainstream media did not welcome President Trump's good news.

A *Washington Post* Editorial Board opinion four days later was illustrative. Under the headline "Trump is spreading false hope for a virus cure—and that's not the only damage," the editors opined:

> No one is sure how it might work against viruses. . . . The most promising answer to the pandemic will be a vaccine, and researchers are racing to develop one. . . . His comments are raising false hopes. Rather than roll the dice on an unproven therapy, let's deposit our trust in the scientists.[43]

Critical readers wondered why the *Washington Post* editors were so confident in their dismissal of Professor Raoult's study. Somehow it seemed they already knew—or thought they knew—that Raoult's findings didn't merit consideration. They further asserted that "only a set of controlled clinical trials" could validate his study, but instead of exhorting President Trump to order them, they proclaimed that true promise was to be found in a vaccine, on which "the scientists" were working (implying that Raoult wasn't a scientist).

How did the editors know this? Decades of research and billions of dollars hadn't yielded a vaccine against HIV or Ebola. Seasonal flu vaccines had always been hit-or-miss and never eradicated influenza viruses. No effective vaccine against coronaviruses had ever been developed, despite multiple attempts. What gave the Editorial Board such confidence that the NIAID-Moderna vaccine, which had only just commenced Phase I human trials, was the solution?

A few days after the *WP* editors published this opinion, they published an opinion by Bill Gates,[44] who asserted that "we need a consistent nationwide policy to shutting down." He concluded his opinion with the assertion that "to bring the disease to an end, we'll need a safe and effective vaccine." One week later, on April 6, he stated in an interview with Trevor Noah on the *Daily Show* that "a vaccine is the only thing that will allow us to return to normal."[45] He also announced that his foundation was going to invest billions in building factories to make COVID-19 vaccines, even before seeing conclusive data on their efficacy. It was critically important, he claimed, to scale up manufacturing *during* testing instead of waiting for trial results. Mark Suzman, CEO of the Gates Foundation, also gave an interview on CNN in which he stated it was imperative to fast-track vaccine development.

"We then need to be ready to scale and manufacture that vaccine in literally hundreds of millions and billions of doses because it needs to go global when we have it," he explained. Gates and his CEO were committed to lightning-fast development and global rollout of a new COVID-19 vaccines, and they were willing to invest billions to make it happen. The labs developing the vaccines had no choice but to succeed in quickly making safe and effective products.

On April 30, Gates wrote on his blog (GatesNotes) that the drug candidates for treating COVID were not powerful enough. He did not name these candidates but included a hyperlink to a report on remdesivir. Drugs like remdesivir could, he claimed, "save a lot of lives, but they aren't enough to get us back to normal. Which leaves us with a vaccine." The world would only be able to go back to normal "when almost every person on the planet has been vaccinated against coronavirus." This assertion laid the foundation for the slogan "A Needle in Every Arm." Vaccinating every person on the planet was going to be a daunting task, but as Gates explained:

> Our foundation is the biggest funder of vaccines in the world, and this effort dwarfs anything we've ever worked on before. It's going to require a global cooperative effort like the world has never seen. But I know it'll get done. There's simply no alternative. . . . We're doing the right things to get a vaccine as quickly as possible. In the meantime, I urge you to continue following the guidelines set by your local authorities.[46]

The message from the man with the checkbook was clear: *There is no treatment apart from remdesivir. Follow the authorities, stay home, stay safe, and wait for the vaccine.*"

CHAPTER 7

The "Simple Country Doctor"

Professor Raoult told the French press that he didn't have months or years to wait for a new therapy or vaccine. Sick patients were arriving at his clinic right then, and he had a duty to treat them. Dr. Vladimir "Zev" Zelenko agreed with him. The family doctor in Monroe, New York, about sixty miles north of New York City, was, like Professor Raoult, a bold and independent spirit. He too had a striking appearance that seemed incongruous with that of conventional medical doctors. An orthodox Jew of the Chabad-Lubavitch movement, he wore a full beard, black kippah, and black suit—a habit that gave outward expression to the fact that he did not live or think in the mainstream.

His office was located near Kiryas Joel—a Hasidic Jewish village of 1.39 square miles with 33,000 residents. SARS-CoV-2 arrived in March and spread through its dense population. Soon sick patients were calling and showing up at his office and wanting to know what he could do for them. What could he do for them? He figured there had to be *something.*

He called a few doctor friends in New York City. They said that *nothing* could be done. The sick should isolate at home and rest. If they got really sick and could no longer breathe, they should go to the hospital, where they would be intubated and put on ventilators. From a colleague who worked at an ER in the city, Dr. Zelenko learned that the outcome for patients on ventilators was poor. From the data gathered thus far, it seemed that a shocking percentage of them would die on the vent.

Come on, Dr. Zelenko thought. *There's got to be some kind of intervention to prevent hospitalization and death.* He jumped on Google and saw that doctors in South Korea were reporting favorable results with hydroxychloroquine combined with zinc. Searching further, he saw Professor Raoult's announcement that he would treat patients with hydroxychloroquine and azithromycin. He then found an instructional video on YouTube called MedCram Episode 34, posted on March 10 by Roger Seheult, MD, a professor at the University of California, Riverside, School of Medicine.

Professor Seheult explained how zinc impairs an enzyme called RNA-dependent polymerase that is essential for RNA viral replication. The trouble with zinc is that it doesn't easily enter cells, so that another chemical is needed to facilitate this action. Hydroxychloroquine seems to act as an ionophore—an entity that transports ions across the cell membrane—that enables zinc to enter cells in sufficient quantities to inhibit viral replication. He cited the February 13 report in the *Korea Biomedical Review* and hypothesized that South Korea's recommendation of the drug in its official treatment guidelines could explain the country's much lower rate of COVID-19 mortality than Italy's.[47]

At this point, Dr. Zelenko reasoned he had sufficient observational reporting and a solid theoretical basis for *trying* hydroxychloroquine + azithromycin + zinc on his high-risk patients—high risk defined as over the age of fifty and/or suffering from comorbidities that increased their susceptibility to severe COVID-19. As for the safety profile of hydroxychloroquine: The drug had been FDA-approved since 1955 and was well tolerated by people of all ages, including pregnant women and nursing mothers. Because it had been around for so long, its patent had expired. That, and the fact that it's easy to manufacture, made it very cheap.

Dr. Zelenko's guiding principle was that the therapeutic intervention was best started as early as possible. This had long been the standard practice with antiviral medications, which are designed to inhibit the initial, viral replication phase of the infection. If replication could be impaired at the onset of symptoms, it could prevent the disease from progressing in severity. Thus, as soon as the patient presented symptoms of COVID-19, Dr. Zelenko prescribed a five-day regimen of the three substances. In March 2020, confirmatory testing took a week—far too long of a delay.

He would never forget the first time he treated a patient for COVID-19 with his protocol. The seventy-year-old with heart problems seemed likely to get into trouble if nothing was done about it. His son-in-law worked for the local volunteer EMT service, of which Dr. Zelenko was the director. About twelve hours after the prescription was filled, Dr. Zelenko called the patient's son-in-law.

"How's he doing?" Dr. Zelenko asked.

"Much better!"

"Really? Already?"

"Yes, his breathing is easier, and he feels much better."

Wow, I guess this stuff really works, Dr. Zelenko thought. And so it went with the patients that followed. Many reported feeling better within twelve hours of taking the first dose. After receiving positive reports from about a hundred patients, he felt an urgent need to share it with the country's leadership. Thus, on March 21, he made a YouTube video—addressed to President Trump—with a presentation of his protocol and favorable results. He believed his observational data was not only significant unto itself, but also corroborated Professor Raoult's observations in Marseille.

Sixteen hours later, Dr. Zelenko received a call from President Trump's chief of staff, Mark Meadows.

"I believe what I'm observing is significant and could potentially help to win the war," he explained to Meadows, who asked for weekly updates. About a week later, Dr. Zelenko got a call from FDA Commissioner Stephen Hahn, who advised him to share his observational data with the National Institutes of Health (NIH). Hahn gave him an NIH contact, who said the Institutes had no means of evaluating real-world evidence from doctors in the field like him. Dr. Zelenko found this astonishing. It was the equivalent of a general in the Pentagon telling a Marine captain in a combat zone that the supreme command had no means of evaluating situation reports from the field. This was, in a word, *bizarre*. At this point Dr. Zelenko decided to go straight to the top, so he wrote a letter to President Trump.

CHAPTER 8

"My detractors are children!"

Back in France, Professor Raoult and his team got to work on a larger study, and on April 9 he published the abstract. Of the 1,061 patients receiving the combination therapy, 973 (91.7 percent) were cured within 10 days. A poor outcome was observed in 46 patients (4.3 percent). Twenty-five of these required ten days of hospitalization or more but were eventually cured. Ten of these were transferred to the ICU, and 5 died (0.47 percent, 74–95 years old). From these findings, Professor Raoult concluded:

> The HCQ-AZ combination, when started immediately after diagnosis, is a safe and efficient treatment for COVID-19, with a mortality rate of 0.5% in older patients. It avoids worsening and clears virus persistence and contagiosity in most cases.[48]

Several prominent French doctors and virologists proclaimed that Professor Raoult's study had no scientific value because it was nonrandomized, and they claimed it contained other flaws. So began a public debate between what the press called the "Parisian Medical Establishment"—which demanded more evidence of the therapy's efficacy before it could be authorized for general use—and Gandalf of Marseille.

The debate drew attention to what struck many as an astonishing fact. For decades, hydroxychloroquine had been available over the counter in France. However, on November 12, 2019, the National Health Security Agency recommended that it be reclassified as a List II substance, with active ingredients harmful to human health. On January 13, 2020, the Ministry of

Health issued a decree placing hydroxychloroquine on List II, and therefore obtainable by prescription only.[49] The timing of this decree—right as SARS-CoV-2 was rapidly spreading—was conspicuous. At a critical moment when no other treatment was offered, the unfettered right of Frenchmen to try a medication that had long been available to them was taken away.

Much of the French mainstream press took the side of the Parisian Medical Establishment in expressing skepticism about the efficacy (and even safety) of Professor Raoult's therapy. Other elements of the French press and society—what one could describe as "conservative" or "populist"—championed the professor, who characterized his opponents as Groupthink mediocrities. That "conservative" elements of French society championed Raoult—the long-haired, Harley-riding individualist—was a striking cultural development. Whereas the postwar French Left had championed flamboyant characters who rebelled against President De Gaulle's powerful state establishment, now it seemed the left-leaning French press was strictly establishmentarian.

Raoult proclaimed that the Parisian medical establishment was applying the wrong evaluative standard to his work. They wanted to see results of large randomized controlled trials (RCTs) such as those performed by pharmaceutical companies on a *new* drug that had never been used on a human population. They were so preoccupied with doctrinaire issues of research methodology that they were incapable of addressing the urgent problem at hand.

COVID-19 had arrived in France like a military invasion. While Raoult was trying to fight the battle with the *only* promising weapon they had, his detractors insisted the weapon undergo extensive and time-consuming testing before it could be deployed. He pointed out that his critics (especially the virologists) had never treated a single patient. Moreover, his patients were *not* representative of how the disease was affecting most people, for whom COVID-19 was mild and resolved on its own without medical intervention. Most of his patients were older people at higher risk of severe illness, and he wasn't about to place half of them in a control group with a placebo. He considered it unethical to give the sick a placebo when there was already reason to believe that a safe medication could help them.

Back in the United States, the NIH funded a study by the American Veteran's Administration that was published on April 21. The authors stated that it was not a randomized clinical trial and not peer-reviewed. It was, they explained, a retrospective analysis of patients who received hydroxychloroquine, and those who didn't, in US Veterans medical centers until

April. They reported that 158 patients received no hydroxychloroquine or azithromycin, while 97 patients received HCQ and 113 patients received the combination therapy advocated by Professor Raoult. Of those who received neither drug, 11 percent died. Of those who got HCQ, 28 percent died; of those who got HCQ and azithromycin, 22 percent died.[50]

CNN's Anderson Cooper quickly publicized this study.

"The President had been out there touting hydroxychloroquine saying, 'What do you have to lose?'"[51] Cooper proclaimed. He then pointed to the VA study's higher death rate for the patients who received hydroxychloroquine. What neither Cooper nor CNN's medical commentator, Sanjay Gupta, mentioned was that the hospitalized patients who received the therapy were much sicker than those who didn't and were given the treatment as a last resort. This decisive difference between the experimental and control groups was ascertained by Professor Raoult when he analyzed the study's tables and noticed that the deceased patients had pathologically low white blood cell counts, signaling they were much closer to death than the patients who didn't receive the therapy. Subsequent white blood cell studies in gravely ill COVID-19 patients confirmed Professor Raoult's observation.[52] Giving a drug to a dying patient as a last resort, and then claiming that his death was causally linked to the drug, is a distortion known as "confounding by indication."

On May 9, Professor Raoult gave an interview with the popular *Paris Match* magazine in which he took the gloves off.

"My detractors are children! . . . I refuse to debate with people whose level of knowledge is so low!"[53] He then proclaimed his treatment protocol to be a no-brainer, which was precisely why it irritated the Parisian medical establishment, with its pharmaceutical industry doctrine:

> We are in a situation where two drugs, available, inexpensive, prescribed billions of times, work. Two solutions present themselves: either you use them, or you try to unearth a hypothetical drug, no more effective, which will cost more and which is not yet manufactured. It's crazy to imagine progress by refusing to use old molecules. Our approach, the same as that of the Chinese, is of biblical simplicity. We test, it works, we prescribe.

To the question of why he suddenly withdrew from the national Scientific Counsel, he explained:

"These people didn't know what they were talking about. . . . There is nothing scientifically reliable about it. . . . In 2003, I wrote a report on epidemic risks, based on my observations on the Chinese reaction to the SARS epidemic. Here, in twenty years, they have learned *nothing*."

The final question of the interview was if he was bothered by the fact that much of the French medical establishment had rejected him.

"Not at all! . . . Remember the last words of *The Stranger*, by Camus: "For me to feel less alone, I wished that there would be many spectators on the day of my execution to greet me with cries of hatred."

In contrast to Dr. Yves Lévy, Professor Raoult was terrible at the politics of medicine. His total lack of diplomacy certainly didn't help his cause. McCullough later became acquainted with the professor via WebEx and was invited to give Grand Rounds for his hospital's academic division. Raoult had a marvelously charismatic air about him. McCullough never heard him utter a word of English to any of his fellow countrymen, so it was surprising how eloquently he spoke the language in private conversations with the American.

McCullough figured Raoult could have penetrated the obtuse American media if he'd taken the time and effort to do so. Factually speaking, there could be no doubt he was right about early treatment. His combination therapy produced a large enough signal of benefit to justify immediately administering it to high-risk patients. The alternative—to do *nothing* to prevent severe illness—was illogical, scientifically unsound, and immoral.

CHAPTER 9

Memento Mori

Nine days after Professor Raoult's colorful interview with *Paris Match*, President Trump gave his daily press briefing, in which he announced that he himself was taking hydroxychloroquine to prevent COVID-19 infection. The assembled reporters responded with shocked incredulity. It was as though the president had just announced he was taking a dangerous experimental drug, and *not* a WHO Essential Medication, taken by millions for decades. Shortly after he made this announcement, Fox News host Neil Cavuto assured his audience, "If you are in a risky population here, and you are taking this as a preventative treatment to ward off the virus . . . it will kill you. I cannot stress enough: this will kill you."[54]

President Trump's proposition that hydroxychloroquine could work as a prophylaxis against COVID-19 was, at this time, being tested by Drs. McCullough and Wheelan at the Baylor University Medical Center. The proposition was especially alarming to the Bio-Pharmaceutical Complex, because one of the statutory requirements for obtaining an FDA Emergency Use Authorization for a new experimental drug is that "there are no adequate, approved, and available alternatives."[55] Because hydroxychloroquine was already FDA-approved for other conditions and had a well-established safety profile, its efficacy in preventing and treating COVID-19 would likely raise questions about an EUA approval for the new mRNA vaccines being developed by Pfizer and Moderna. This was a key motive for smearing hydroxychloroquine in the mainstream media.

The reporters at President Trump's May 18 press conference asked him why he was taking it.

"Because I think it's good and I've heard a lot of good stories about it," he said:

> And if it's not good, I'm not going to get hurt by it. It's been around for forty years for malaria, for lupus, for other things. Frontline workers take it, a lot of doctors take it, I take it, and I think people should be allowed to. I got a letter from a doctor the other day in the Westchester, New York, area. He didn't want anything. He just said, "Sir, I have hundreds of patients. And I give them hydroxychloroquine, I give them the Z-Pack, which is azithromycin, and I give them zinc. And out of the hundreds of patients, many hundreds, over three hundred patients, and I haven't lost a patient."[56]

It was quite a milestone in Dr. Zelenko's career. Just six weeks earlier he'd lived in perfect obscurity. Now the president of the United States was referring to his letter and endorsing his findings on national television. To Dr. Zelenko, it was a surreal experience, and the irony wasn't lost on him that in a nation with the largest academic medical establishment on Earth, it wasn't an eminent professor of medicine, but a small-town family doctor who advised the president.

Throughout history, in every field of science and technology, there has been a tension between practical experimentation and academic authority. Dr. Zelenko's experience recalls that of the English clockmaker John Harrison, who found a practical solution to a massive problem that had, since ancient times, plagued open-ocean navigators—that is, how to determine a vessel's longitude. This wasn't an academic problem. Over the centuries, countless ships had run aground or been lost at sea because their navigators couldn't determine their longitudinal location. Thus, the solution was a matter of life or death for thousands of mariners.

For John Harrison, the solution was to build a clock that would keep perfectly accurate time at sea, regardless of the ship's motion and changes in humidity. The difference between the Greenwich Mean Time kept on the clock and local time at noon fixed the vessel's longitude. Many of the great academic mathematicians and astronomers of his day regarded his project as absurd. Apparently to prevent Harrison from winning the large sum of prize money offered by Parliament to whoever solved the longitude problem, the Astronomer Royal, Nevil Maskelyne, used his position of authority to suppress the truth that Mr. Harrison's clock was, in fact, the solution. It was only after Harrison succeeded in getting an audience with "Mad" King George III that his marine chronometer got the recognition it deserved and ultimately became standard navigational equipment.[57]

For Dr. Zelenko, treating COVID-19 wasn't only a matter of gauging outcomes. In his view, a doctor had a sacred duty *to try* to help his sick patients, even with no guarantee of success. Writing an off-label prescription for an FDA-approved medicine had long been a legal and common practice. The original drug company advertising claims in the label have no bearing on the evidence-based use of drugs in the present time. In fact, off-label prescribing was explicitly endorsed by the FDA in its 2018 guidance on unmet medical needs.[58]

Dr. Zelenko also knew from personal experience that—when faced with a probable death sentence—being able *to try* a therapy is an inalienable human right. Two years earlier he'd been diagnosed with a rare form of cancer that, in almost all cases, is only discovered in postmortem examinations. In his case, the tumor was detected and removed before it killed him. From his experience of undergoing two open-chest surgeries to treat a fatal and spreading disease, Dr. Zelenko knew how little time we have. Living under a death sentence prompted him to conduct his own research. He found a reference to an experimental drug therapy that seemed to offer a glimmer of hope. He booked a consultation with an oncologist in New York City to discuss it.

"Did you know I'm one of the authors of that study?" the doctor said.

"No, I didn't see your name on it," Dr. Zelenko replied. There was no way to know for sure if the combination therapy was the decisive factor, but something seemed to be holding the cancer at bay, because two years later, he was still alive and kicking. If a patient is certain or likely to die without an intervention, it was only logical that he would try a therapy that seemed to offer some hope, even if its efficacy wasn't fully established. For the state to prohibit this right to try was not only irrational, but tyrannical.

Tyranny and its long, dreadful history were familiar to Dr. Zelenko, who was born in Kiev, Ukraine, when it was still under Soviet control. One of his favorite books was Aleksandr Solzhenitsyn's *The Gulag Archipelago*—a meticulous and terrifying account of how a society lapses into tyranny, with individuals making hundreds of compromises in order to get along, only to one day find themselves in a gulag. Dr. Zelenko saw the same spirit moving through America's medical complex, and it filled him with dread—not for himself, but for his family and country.

A positive consequence of living under a death sentence was that it liberated him from fear of other human beings. Having come to terms with his death, the possibility of encountering dangerous and malevolent people ceased to scare him. Thus, in a strange way, dying from metastatic sarcoma had spiritually prepared him for the battle that began in March 2020. He joined it knowing that the time to do the right thing is *now*. The young and strong assumed they could put it off till later, but this was an illusion.

The lesson that Dr. Zelenko learned from his cancer diagnosis had long been understood by Professor Raoult. A student of Greek and Roman Stoic philosophy, he too knew that none of us have time to procrastinate the business of facing life's challenges and savoring them. On his right pinkie he wore a silver ring bearing the likeness of a skull. Many assumed it was a "biker ring" that complemented his Harley-Davidson. In fact, it was a memento mori—a reminder of death. Being mindful of it kept him focused on the essential and freed him from sweating the small stuff. Like Dr. Zelenko, he didn't shy from a fight. On the contrary, he relished it.

By all appearances, the president of the United States was on their side, but was Donald Trump the right man for the job?

CHAPTER 10

Shooting the Message

Long before COVID arrived, I'd joked with friends that if the ever-boasting and self-promoting President Trump discovered the cure for cancer, his detractors would rather forgo the treatment than acknowledge he'd done something of value for humanity. It's probably an inevitable outcome of America's two-party system that any president may become the object of hyperbolic loathing. In the nineties, Republicans obsessed about President Clinton's faults, real and perceived. In 2003, the columnist and psychiatrist Charles Krauthammer coined the expression Bush Derangement Syndrome as "the acute onset of paranoia in otherwise normal people in reaction to the policies, the presidency—nay—the very existence of George W. Bush."

During the presidency of Donald J. Trump this syndrome became so virulent that it created a strangely binary posture in public affairs. If Trump expressed even mild enthusiasm for a policy, person, or thing, his opposition automatically rejected it. To be sure, Trump often threw gasoline on the fire with his vices, his bombastic personal style, and occasional buffoonery. The qualities that had once been viewed as showman's shtick were widely deemed unacceptable in a US president. The court jester had become king, and it drove the lords and ladies at court mad.

As anthropologists and psychologists have long understood, humans are hypersocial and tribal. Stanford Professor Renée Girard has pointed out that during times of stress and rivalry, we are inclined to ascribe blame not to a complex state of affairs, but to a particular person or group. Persistent problems and misfortunes build up negative psychic energy, which generates a collective yearning to destroy the person or persons on whom the blame is heaped. This process of *scapegoating* is amplified by what Professor Girard called *mimesis*, that is, imitation, the tendency to embrace an opinion or sentiment because everyone in a preferred group is embracing it. In trying to make sense of the world, we often look to those around us for cues to guide us in our perceptions and opinions.

So, it was when President Trump declared the old malaria drug hydroxychloroquine to be a potential game changer. Soon stories appeared under headlines such as "Trump's COVID Cure" or "Trump touts hydroxychloroquine," proclaiming the drug lacked efficacy and safety and caused "irreversible retinal damage" and "dangerous heart arrhythmias."

As any ophthalmologist could tell you, the claim of "irreversible retinal damage" from a five-day course of hydroxychloroquine was a whopper of a deception. Less than 1 percent of lupus and rheumatoid arthritis patients who take hydroxychloroquine every day *for five to seven years* start to develop retinal toxicity.[59] At ten years of daily use, 1 percent of regular users develop retinal toxicity. Likewise, the claim of "dangerous heart arrhythmias" was a massive distortion.

On April 7, Dr. Oz interviewed the prominent rheumatologist Dr. Daniel Wallace, who specialized in treating systemic lupus at Cedars-Sinai Medical Center in Beverly Hills. Dr. Oz began the interview by mentioning that 400,000 Americans are currently using the drug.

"How safe is it?" Dr. Oz asked:

> In forty-two years of clinical practice, I've treated several thousand lupus patients, and I'd like to emphasize that all rheumatologists have a great deal of experience with this drug. Regarding safety, since it came out seventy years ago in 1955, several million patients have taken the drug. There have not been any reported deaths from using this agent as a monotherapy . . . It [heart arrythmias] was a problem with hydroxychloroquine in the nineteen fifties and sixties, when doctors were using two to three times the usual dose. In the current recommended dose, it really does not occur. . . . The risk of retinal toxicity in five years of continuous use is zero.[60]

Dr. Wallace was only speaking about hydroxychloroquine as a monotherapy. The incidence of dangerous arrhythmias for patients taking hydroxychloroquine *plus azithromycin* is approximately 9 in 100,000.[61]

Another big lie frequently told in the press was that "Trump's touting" of hydroxychloroquine was hindering lupus and rheumatoid arthritis patients from getting their prescriptions filled. The generic drug is easy to manufacture, and production could have easily been ramped up. Moreover, in late March, the White House Office of Trade and Manufacturing Policy (OTMP) acquired 29 million tablets of hydroxychloroquine for the U.S. stockpile.[62] If there was indeed a supply chain problem, it was created by meddling officers at the Department of Health and Human Services (HHS) and the FDA, who obstructed the national stockpile from being distributed to pharmacies.

Many marveled at the drug's hostile politicization, but it's important to understand that inciting and manipulating political passions was only a means to an end. The objective was to suppress hydroxychloroquine as a treatment for COVID-19. The motive was revealed in the *Washington Post* editorial of March 25, 2020. Already then, the DC medical establishment had decided that a vaccine was the answer to the COVID-19 pandemic. No one in the government apart from President Trump was speaking favorably about repurposing drugs such as hydroxychloroquine. The widespread loathing of Trump was not the motive for discrediting the therapy, but it helped public health agency directors to turn the public against it.

A few months after Dr. Zelenko wrote his letter to President Trump, he learned that a ranking officer at the HHS sabotaged the president's executive order to make hydroxychloroquine readily available. This revelation came in October 2020 with the release of the documentary film *Totally Under Control,* for which Dr. Zelenko was interviewed. Also interviewed was Dr. Rick Bright, who in March 2020 was the director of the Biomedical Advanced Research and Development Authority (BARDA)—an HHS branch founded in 2006 and engaged in "the development of the necessary vaccines, drugs, therapies, and diagnostic tools for public health medical emergencies." The Authority is especially interested in vaccine development and was—along with Anthony Fauci's NIAID—a major funder of Moderna's mRNA vaccine against COVID-19.

President Obama appointed Dr. Bright to direct BARDA in 2016. According to his CV, he has also "served as a key advisor in a number of roles, including in the development of the Coalition for Epidemic Preparedness Innovations (CEPI), the WHO Research and Development Blueprint for Action to Prevent Epidemics, the WHO Global Action Plan for Influenza vaccines."[63] On October 29, 2019, Dr. Bright (along with Dr. Fauci) attended a conference at the Milken Institute to discuss the need for a universal flu vaccine. The moderator, *New Yorker* staff writer Michael Specter, repeatedly lamented that there wasn't sufficient motivation to invest the resources necessary to make the "disruptive" leap from traditional vaccine technology to new vaccine technology. The trouble was, people only seemed to get motivated to make the investment after a new pandemic emerged. At one point he posed a provocative question.

"In the long run, over time, amortized, if the 2009 pandemic had been much more deadly, would that have ended up being a better thing for humanity?" With this question, Mr. Specter fell into the mental trap of 20th-century dictators in thinking of "humanity" in the abstract instead of as individual, suffering men and women. To his credit, Dr. Fauci answered no, though his reasoning was also utilitarian.

"We had a pretty bad pandemic in 1957 and 1968, and that didn't change much," he answered.

"But don't we have some biotechnological tools now at our disposal that we didn't then?" Specter replied. This initiated a conversation about these new "biotechnological tools." Dr. Bright opined that the most promising new technology to combat the flu or other viral pandemics was "nucleic acid-based, and messenger RNA based sequences that can be rapidly shared around the world."[64]

In his interview for *Totally Under Control*, Dr. Bright explained how he thwarted President Trump's executive order to make the national stockpile of hydroxychloroquine widely available to the public. He initiated this scheme on March 23, when HHS Secretary Alex Azar's chief counsel directed him to take aggressive action to distribute hydroxychloroquine from the national stockpile to pharmacies. The next day, Dr. Bright spoke with Dr. Janet Woodcock (FDA Director of the Center for Drug Evaluation and Research), and they decided the best way to undermine the president's order was for Dr. Bright to submit an Emergency Use Authorization (EUA) request for hydroxychloroquine, but with the condition that the drug be restricted to hospitalized patients.[65]

Dr. Bright apparently assumed that hydroxychloroquine was potentially dangerous for people in the early outpatient setting who wished to prevent hospitalization, but not particularly dangerous for gravely ill people in hospital. He offered no scientific or logical reason for making this assumption. As Dr. Stephen J. Hatfill remarked in a critical review of the incident, Dr. Woodcock's participation in the scheme was even more puzzling. Unlike Dr. Bright, a PhD with no clinical medical training or experience, she was an internal medicine specialist with subspecialty training in rheumatology. Thus, she certainly knew the top safety profile of hydroxychloroquine as it had been administered to lupus, scleroderma, and rheumatoid arthritis patients for over fifty years, often in much higher doses for much longer times than the proposed dosing for early COVID-19 cases.[66]

In *Totally Under Control*, Dr. Bright claimed he "was proud of our team for coming up with that compromise that we thought would protect Americans." The FDA granted Dr. Bright's request on March 29, which gave many doctors and pharmacy boards the impression they weren't authorized to write and fill prescriptions for nonhospitalized patients. Lupus and rheumatoid arthritis patients could still receive it, but not COVID-19 patients.

As Dr. Bright further explained, six days after the EUA was issued, he woke up to an e-mail with multiple HHS, FEMA, and FDA recipients, stating:

> WH call. Really want to flood NY and NJ with treatment courses. Hospitals have it. Sick out-patients don't. And can't get. So go through distribution channels as we discussed. If we have 29 million doses, send a few million ASAP. WH wants follow up in AM.

At this point in his interview, Dr. Bright's face darkens, and his voice is imbued with indignation:

> And in that e-mail discussion, someone was brave enough to say, "Wait a minute, the EUA says for hospitalized patients only." The reply was "No, it needs to go to the pharmacies, the EUA doesn't matter, push the drug out now." I lost it! I lost all respect for that chain of command, for that security that we, as Americans, have put in those people. And that was the moment I decided to break protocol.[67]

He broke protocol by contacting media outlets and telling them that President Trump was recklessly pushing "the unproven treatment" hydroxychloroquine. Dr. Bright's objective was, he claimed, to protect Americans and to insist that the "government invest the billions of dollars allocated by Congress to address the COVID-19 pandemic into safe and scientifically vetted solutions, and not in drugs, vaccines, and other technologies that lack scientific merit."[68]

Dr. Zelenko regarded Dr. Bright's self-congratulatory statements as a confession of mass homicide. The whole point of the therapy was to administer it early to *prevent* hospitalization. Once the patient landed in hospital, it was probably too late for the drug combination to work. Dr. Bright's histrionic statement is a glaring example of medicine being politicized. By his own admission, he and his scientific team at BARDA had no idea if hydroxychloroquine worked against COVID or not. He *claimed* his team could find no scientific rationale for using the drug, even though research teams in China, South Korea, and France had, by then, reported favorable results with it.

In the United States, Dr. Robert Malone obtained reports on Chinese studies of hydroxychloroquine's efficacy in February 2020 and passed these to his contacts at the CIA and the HHS's assistant secretary for Preparedness and Response. Thus, as Dr. Malone stated in his December 31, 2021, interview with Joe Rogan, "The assertion that there was no data on hydroxychloroquine's efficacy at the time [that Rick Bright made his determination] is patently false."[69]

Dr. Bright claimed his actions were to protect the American people from hydroxychloroquine, but he cited no evidence the medication was harmful. He implied that President Trump's executive order was plainly reckless—a message the president's detractors in the legacy media were eager to hear. He glossed over the fact that delivering the drug to New York and New Jersey pharmacies did not alter the authority of licensed medical doctors to write the prescriptions in accordance with their judgment.

An arrogant presumption was implied in Dr. Bright's statements that only he and his team at BARDA could make a sensible decision about the matter. And even though he lacked legal authority to do so, he took it upon himself to impede Americans from accessing the drug. About a week after he "broke protocol" and directed a media campaign against hydroxychloroquine, he oversaw the transfer of $483 million from BARDA to Moderna to develop its mRNA vaccine.[70]

Later Dr. Bright was demoted within the HHS, to which he responded with a widely publicized whistleblower complaint against the Trump Administration for what he claimed was an unlawful retaliation against him. This gave him another opportunity to talk with the media about his brave protest of the president's "political interference over science and the spread of inaccurate information that was 'dangerous, reckless and causing lives to be lost.'"[71] It also gave him another occasion "to insist that the government invest the billions of dollars allocated by Congress to address the COVID-19 pandemic into safe and scientifically vetted solutions."[72] He then left BARDA to become CEO of the Rockefeller Foundation's Pandemic Prevention Institute and senior vice president of its Pandemic Prevention & Response.

A couple of months later, McCullough wrote a letter to FDA commissioner Dr. Stephen Hahn. The purpose of his letter was to support a request—submitted by Drs. William O'Neill, John McKinnon, Dee Dee Wang, and Marcus Zervos at the Henry Ford Health Center—for an EUA for hydroxychloroquine for "ambulatory prophylaxis and treatment of COVID-19." These doctors had just performed a major study of the drug's efficacy and safety, with very favorable results. McCullough, who'd previously worked at Henry Ford Health, knew the doctors well and vouched for their competence. He also mentioned the promising preliminary results of his Baylor prophylaxis study, with no adverse events related to the medication.

Shortly after he sent his letter to Commissioner Hahn, he received a call from Peter Navarro, director of the Office of Trade and Manufacturing Policy. Mr. Navarro wanted to speak directly with McCullough about the safety and efficacy of hydroxychloroquine. Quickly it became apparent that he was staunchly opposed to how BARDA and the FDA had handled the drug. He spoke passionately, in very colorful language with some of the tightest groupings of F-bombs that McCullough had ever heard.

McCullough reiterated what he'd stated in his letter to Commissioner Hahn. Hydroxychloroquine's safety was well characterized from millions of doses taken since 1955. It worked against malaria through multiple mechanisms of action that seemed to inhibit the cellular invasion and replication of SARS-CoV-2, especially when combined with azithromycin and zinc. The results of the Henry Ford study were indeed very favorable. The totality of circumstances indicated the drug should be available for ambulatory prophylaxis and treatment of COVID-19. Mr. Navarro seemed encouraged by McCullough's assurances and said he would do what he could to persuade the FDA to issue the EUA.

CHAPTER 11

"Cuomosexuals"

On the same day (March 23) that Dr. Bright initiated his scheme to restrict hydroxychloroquine to hospitalized patients, New York Governor Andrew Cuomo issued the following executive order:

> No pharmacist shall dispense hydroxychloroquine or chloroquine except when written as prescribed as an FDA-approved indication; or as part of a state approved clinical trial related to COVID-19 for a patient who has tested positive for COVID-19 with such test result documented as part of the prescription. No other experimental or prophylactic use shall be permitted . . . [73]

This order prohibited New York pharmacies from filling off-label prescriptions for COVID patients. The exceptionally determined Dr. Zelenko found a way to get around it, but it made his practice much more difficult.

Two days later, on March 25, the New York Department of Health issued the following directive to nursing home administrators:

> No resident shall be denied re-admission or admission to the NH solely based on a confirmed or suspected diagnosis of COVID-19. NHs are prohibited from requiring a hospitalized resident who is determined medically stable to be tested for COVID-19 prior to admission or readmission.[74]

As anyone who has ever worked in a nursing home knows, respiratory viruses can rip through the facility and cause severe illness. Virulent common cold rhinovirus outbreaks have resulted in multiple deaths in this setting.[75] By

March 25, it was crystal clear that the risk of severe disease and death from COVID-19 is by far the highest for patients over the age of seventy-five.[76] Thus, ordering "confirmed or suspected" COVID patients to be readmitted to nursing homes was the equivalent of forcing foxes into henhouses. What was the New York State Health Department thinking?

On March 27, the United States set the world record of COVID-19 cases, and New York City was the nation's epicenter—a five-alarm fire of serious infections. On April 10, New York State recorded more COVID cases than any country on Earth except the United States in total,[77] and its nursing homes were devastated by the contagion. The legacy media were slow to notice this, perhaps because they were blinded by Governor Cuomo's stardom. He gave daily press conferences in which he spoke about the measures he and his team were taking to keep New Yorkers safe. Millions across the country watched these performances for which he later received an Emmy.

As the spring wore on, reports of mass casualty events in nursing homes emerged, and these drew scrutiny to the Health Department's March 25 directive. On May 21, the Associated Press reported that over 4,500 COVID patients had been sent back into New York nursing homes.[78] This number would later be revised upward to over 9,000.[79] The AP report coincided with growing suspicion the Health Department wasn't being transparent about mortality data in these facilities. Suspicion was confirmed on January 28, 2021, when New York Attorney General Letitia James reported that the Department of Health had undercounted nursing home deaths by 50 percent.[80]

On February 11, 2021, the *New York Post* published a leaked audio recording of Governor Cuomo's secretary, Melissa DeRosa, speaking confidentially with the New York State Democratic Committee. On this tape she can be heard apologizing for concealing nursing home data. Though mealy-mouthed, her apology revealed that Cuomo's team had acted out of fear of getting into trouble with the DOJ:

> Basically, we froze because then we were in a position where we weren't sure if what we were going to give to the Department of Justice or what we give to you guys and what we start saying was going to be used against us, and we weren't sure if there was going to be an investigation.[81]

After making this confession, she changed the subject to "the context" of their decision—namely, they were concerned that President Trump would try to politicize the incident in the upcoming election.

"Right around the same time, he [Trump] turns this into a giant political football," she told the Committee.

For many who followed the New York nursing home story, it seemed emblematic of many pathologies afflicting the US political and media class. First was the nepotism of the Cuomo family, with CNN Anchor Chris and Governor Andrew regularly putting on shows for their fawning, sentimental fans, many of whom called themselves "Cuomosexuals."[82] T-shirts, hats, coffee mugs, and even a popular music video appeared bearing the term's definition: *In love with competent, reassuring governance by a leader who uses complete sentences and displays common sense during a pandemic.*

The Governor's cult of personality yielded a $5.2 million book deal with Penguin Random House, initiated by an acquiring editor on March 19, 2020, three days before the state went into lockdown.[83] The deal for *American Crisis: Leadership Lessons from the COVID-19 Pandemic* stipulated the book be ready for publication before the November elections. Governor Cuomo purportedly wrote a 70,000-word typescript in three months while at the same time executing his duties as full-time "Leader" in handling New York's COVID crisis. The state ethics board approved the deal on the condition that no state resources be used in the book's production, but that didn't stop Cuomo from using his staff and a ghostwriter.[84]

Complementing the governor's book deal was his Emmy Award. As Bruce Paisner, CEO of the International Academy of Television Arts and Sciences, explained in his announcement of November 20, 2020:

> The Governor's 111 daily briefings worked so well because he effectively created television shows, with characters, plot lines, and stories of success and failure. People around the world tuned in to find out what was going on, and *New York tough* became a symbol of the determination to fight back.[85]

All these awards and extravagant expressions of adulation for the governor's leadership overlooked his executive order impeding access to hydroxychloroquine and his Health Department's catastrophic directive to nursing homes. Protecting nursing homes was the *only* contagion control policy for which there was a crystal-clear rationale. While general lockdowns did little to stop the spread, extraordinary measures to secure nursing homes probably would have given some protection to society's most vulnerable. Instead, the New York Health Department sent thousands of COVID patients back into these facilities and then concealed the ensuing death toll. On June 2, 2020, *USA Today* reported that "Over the last three months, more than 40,600 long-term care residents and workers have died of COVID-19—about 40% of the nation's death toll attributed to the coronavirus . . ."[86]

After flying high in 2020, the Cuomo brothers fell back to Earth in 2021, when multiple women accused the governor of sexual harassment. He was then further accused of using his executive power to suppress these allegations.[87] Chris Cuomo was likewise accused of using his powerful media connections to aid and abet his brother in the concealment.[88]

A cynic might be tempted to wonder about the timing of the sexual misconduct allegations—right as reports emerged that New York Attorney General Letitia James, U.S. Attorney Seth Ducharme of the Eastern District of New York, and the FBI were opening investigations into allegations of malfeasance resulting in nursing home deaths. Especially disturbing was the allegation that Governor Cuomo provided legal immunity to nursing home executives from whom he received campaign contributions, possibly giving them carte blanche to cut costs at the expense of the care and safety of their residents. As the attorney general stated in her preliminary findings:

> On March 23, Governor Cuomo created limited immunity provisions for health care providers relating to COVID-19. The Emergency Disaster Treatment Protection Act (EDTPA) provides immunity to health care professionals from potential liability arising from certain decisions, actions and/or omissions related to the care of individuals during the COVID-19 pandemic. While it is reasonable to provide some protections for health care workers making impossible health care decisions in good faith during an unprecedented public health crisis, it would not be appropriate or just for nursing homes owners to interpret this action as providing blanket immunity for causing harm to residents.[89]

With multiple allegations of sexual misconduct made in March 2021, the subject in mainstream media reporting was largely changed from New York State's possible liability for the preventable deaths of thousands to Governor Cuomo's inappropriate behavior with women.

On August 7, 2021, the *New Yorker* published a coda to Governor Cuomo's rise to superstardom and his crashing fall from grace. In an essay titled *Diving Into the Subconscious of the "Cuomosexual,"* reporter Lizzie Widdicombe posed the question: *How could we have witnessed the Governor's narcissism, bullying, and hackneyed paternalism, and found these qualities attractive?* To answer it, she interviewed psychoanalyst Virginia Goldner, who explained that Governor Cuomo "was radiating an eroticized masculinity that has within it hostility and a little tenderness. That combination of soft and hard—mostly hard, but also soft—is what so many women crave in some way."[90]

Dr. Goldner's remarks pointed to a key feature of how the public responded to official COVID policy. Approval of policies had little to do with their substance. Mostly it derived from impressions of the personal qualities, political affiliation, and perceived authority of the officials who presented the policies. Governor Cuomo exuded masculine confidence and gave the impression of taking bold action against a foreign invader. His performances were fascinating to watch, but they had little to do with reality.

By late March of 2020, the virus had spread far beyond the possibility of being contained. The Swedish state epidemiologist, Anders Tegnell, pointed this out in a March 28, 2020, *New York Times* interview,[91] but no major public health official in the United States acknowledged this reality. Because the virus was far beyond containment, it was unlikely that any of Governor Cuomo's contagion control orders such as his statewide lockdown or shutdown of "nonessential businesses" made any positive difference. He was awarded an Emmy for embodying "the determination to fight back" against the virus. In fact, he disarmed New Yorkers by impeding their access to the only weapon (hydroxychloroquine) known at the time for fighting it. COVID patients, including thousands of sitting ducks in nursing homes, were consequently left defenseless.

CHAPTER 12

The Wonder Drug

On April 3, 2020, while the media was fixated on President Trump's endorsement of hydroxychloroquine and Governor Cuomo's performances, a research team at the Monash University of Australia published a study on the antiparasitic drug ivermectin. The team concluded:

> Ivermectin is an inhibitor of SARS-CoV-2 virus in vitro. A single treatment effected approximately 5000-fold reduction in virus at 48 h in cell culture. Ivermectin is FDA-approved for parasitic infections and included on the WHO model list of essential medicines, and is thus widely available.[92]

It would be hard to overstate the significance of this discovery. Like penicillin, derived from the Penicillium mold, ivermectin is a bioactive compound of natural origin, derived from the Streptomyces avermectinius bacterium. The discovery of this bacterium and its derivative drug, ivermectin, is one of the most fascinating in medical history. Since its large-scale human distribution began in 1989, ivermectin has cured two great scourges that had, for centuries, afflicted millions of people in the tropics—Onchocerciasis (commonly known as River Blindness) and Lymphatic filariasis (commonly known as Elephantiasis).[93]

The Streptomyces avermectinius bacterium was discovered by the Japanese biochemist Satoshi Ōmura in woodland soil near a golf course on the southeast coast of the island of Honshu, Japan. Dr. Omura isolated the bacterium from this soil sample—the only in which it has ever been found—and sent it to his colleague William Campbell at the Merck

Institute of Research in the United States. From this bacterium, Campbell derived the compound macrocyclic lactone, to which he gave the name ivermectin. Quickly Campbell determined it was a potent killer of parasitic roundworms with virtually no toxicity for animals, and in the early eighties, ivermectin became a blockbuster in veterinary medicine.

Another Merck scientist, Mohammed Aziz, then worked with the World Health Organization to test ivermectin against River Blindness in humans. They found the drug to be extremely safe and effective, prompting Merck to initiate a mass donation program to countries in which the disease was endemic. Ivermectin thereby earned the rare title of "wonder drug," protecting millions of people from River Blindness and Elephantiasis.[94]

For their spectacular contribution to human health, Satoshi Ōmura and William Campbell received the Nobel Prize in Medicine in 2015,[95] but the story of ivermectin didn't end there. The February 15, 2017, issue of *The Journal of Antibiotics* featured a report titled "Ivermectin: enigmatic multifaceted 'wonder' drug continues to surprise and exceed expectations." The report presented ivermectin's therapeutic properties against an array of pathogens, including viruses:

> Ivermectin has been found to potently inhibit replication of the yellow fever virus. . . . It also inhibits replication in several other flaviviruses, including dengue, Japanese encephalitis and tick-borne encephalitis. . . . Ivermectin inhibits dengue viruses and interrupts virus replication, bestowing protection against infection with all distinct virus serotypes, and has unexplored potential as a dengue antiviral. . . . Ivermectin has also been demonstrated to be a potent broad-spectrum specific inhibitor of importin α/β-mediated nuclear transport and demonstrates antiviral activity against several RNA viruses by blocking the nuclear trafficking of viral proteins. It has been shown to have potent antiviral action against HIV-1 and dengue viruses . . . [96]

This 2017 report on ivermectin resembled a 2012 Israeli study on hydroxychloroquine's array of benefits and unexplored potential. Both medicines—like penicillin and aspirin—derived from compounds found in nature. Many of the most useful drugs in history are of natural origin. Eleven percent of the WHO's list of 252 Essential Medicines are found in flowering plants. As pharmacology professor Dr. Ciddi Veeresham put it in a 2012 paper:

> Nature, the master of craftsman of molecules created almost an inexhaustible array of molecular entities. It stands as an infinite resource for drug

> development, novel chemotypes and pharmacophores, and scaffolds for amplification into efficacious drugs for a multitude of disease indications and other valuable bioactive agents.[97]

Dr. Ōmura spent most of his career scouring nature for useful molecules. Ivermectin was just one of many he discovered, though it was certainly the most useful. The cheap generic drug now has a top safety profile, obtained from 250 million people who have taken it annually for decades. No wonder the Monash University research team decided to investigate the possibility that ivermectin had antiviral potential against SARS-CoV-2.

Following the publication of their study on April 3, research teams and independent doctors all over the world studied ivermectin for the treatment of COVID-19, and soon several favorable reports were published. However, as was the case with positive reports on hydroxychloroquine in February and March, the silence of US public and university research institutions about ivermectin's promise was deafening. No hopeful mention of the drug or calls to research it on federal health agency websites or press briefings. Nothing mentioned about it on evening news shows. No op-eds in major newspapers by medical scientists.

The only public health agency that mentioned the April 3 Australian study was the FDA's Center for Veterinary Medicine. On April 10, its Director, Dr. Steven Solomon, issued a *Letter to Stakeholders: Do Not Use Ivermectin Intended for Animals as Treatment for COVID-19 in Humans.* The statement expressed concern about consumers taking veterinary preparations of ivermectin and gave notice that the FDA was investigating fraudulent COVID-19 remedies for possible enforcement action. Regarding ivermectin's apparent promise in treating COVID, Dr. Solomon stated, "Additional testing is needed to determine whether ivermectin might be safe or effective to prevent or treat coronavirus or COVID-19."[98] Implied in this statement was that some researcher might get around to testing ivermectin. However, until his test results were approved by the FDA, the public would just have to accept that there were no treatments for the disease.

A notable exception to the lack of ivermectin research in the US was a study performed by Drs. Jean-Jacques and Juliana Cepelowicz Rajter—a husband-and-wife team who owned an independent pulmonary practice in Florida. They conducted a retrospective cohort study of 280 consecutive patients hospitalized at four Broward Health hospitals with confirmed SARS-CoV-2, of whom 173 were treated with ivermectin and 107 with usual care were reviewed. The study showed "lower mortality in the ivermectin group (15.0 percent versus 25.2 percent). Mortality was also lower among 75 patients with severe pulmonary disease treated with ivermectin (38.8 percent vs 80.7 percent)." This extremely positive study was published in preprint on the medRxiv server on June 6, 2020, and received zero public health agency or media attention. The NIH and BARDA remained silent about it. Later the study was published in the prestigious *CHEST* journal of pulmonary medicine and again received no MSM or health agency attention.[99]

In the months that followed, favorable reports about ivermectin were published all over the world. These included case studies from independent doctors; epidemiological studies in India, Peru, and Mexico; and clinical trials. These showed high safety and significant benefit from ivermectin in COVID patients.[100] Nevertheless, this "Wonder Drug" that had relieved immense suffering from other illnesses and won its discoverers the Nobel Prize in 2015 was ignored by the US Bio-Pharmaceutical Complex until August 27, 2020, when the NIH issued a recommendation *against* using it.[101]

In the six months that followed the NIH's negative advisory, several additional high-quality studies showing ivermectin's efficacy were published. By January 11, 2021, the tally had reached 64 controlled trials—32 randomized, and 16 double-blinded.[102] Nevertheless, the NIH never recommended the drug, and the FDA never approved it for COVID-19—not even under the "compassionate use" provision for sick patients for whom nothing else was working.

CHAPTER 13

Dr. Fauci Goes to Bat for Remdesivir

The NIH insisted there wasn't enough evidence to warrant recommending the old generic drugs hydroxychloroquine and ivermectin for the treatment of COVID-19. This contrasted with its Advisory Committee's remarkably lax evaluation of the new, experimental drug remdesivir, which it recommended in its May 1, 2020, Treatment Guidelines. The NIAID and CDC had a vested interest in remdesivir, as they'd spent $79 million in developing it with Gilead Sciences.[103] Additionally, eleven members of the Advisory Committee had financial relationships with Gilead. Two were on the company's advisory board. Others were paid consultants or received research support or honoraria.[104]

In his book *The Real Anthony Fauci,* Robert F. Kennedy Jr. presented the results of his meticulous and exhaustively documented investigation of remdesivir's curious history:

> In 2016, remdesivir demonstrated middling antiviral properties against Zika, but the disease disappeared before the expensive non-remedy got traction. After the Zika threat vanished, NIAID put some $6.9 million into identifying a new pandemic against which to deploy remdesivir. In 2018, Gilead entered remdesivir in a NIAID-funded clinical trial against Ebola in Africa. . . . Dr. Fauci had another NIAID-incubated drug, ZMapp, in the same clinical trial, testing efficacy against Ebola alongside two experimental monoclonal antibody drugs. . . . However, six months into the Ebola study, the

> trial's Safety Review Board suddenly pulled both remdesivir and ZMapp from the trial. . . . Within 28 days, subjects taking remdesivir had lethal side effects including multiple organ failure, acute kidney failure, septic shock, and hypotension, and 54 percent of the remdesivir group died—the highest mortality rate among the four experimental drugs. . . . NIAID . . . researchers published the bad news about remdesivir in the *New England Journal of Medicine* in December 2019. By then, COVID-19 was already circulating in Wuhan. But two months later, on February 25, 2020, Dr. Fauci announced, with great fanfare, that he was enrolling hospitalized COVID patients in a clinical trial to study remdesivir's efficacy. For important context, this was a month before the WHO declared the new pandemic, a time that there were only fourteen confirmed COVID cases in the United States, most from the Diamond Princess cruise ship. These individuals were among the first wave of COVID-19 hospitalizations from whom NIAID recruited the 400 US volunteers for Dr. Fauci's remdesivir trial. Dr. Fauci's press release said only that remdesivir "has shown promise in animal models for treating Middle East Respiratory Syndrome (MERS)."[105]

Remdesivir wasn't conceived as an early intervention to prevent hospitalization, as it is IV-administered in hospitals. Had the NIH and FDA acknowledged the efficacy of hydroxychloroquine and ivermectin for treating COVID-19, they would have undermined their rationale for granting remdesivir an EUA. General administration of hydroxychloroquine and ivermectin in the early, outpatient setting would have also reduced the number of hospitalizations, and therefore the demand for remdesivir. Again, to quote Robert F. Kennedy Jr.:

> Dr. Fauci did not suddenly get the idea that remdesivir might work against coronavirus in January 2020. In one of his many extraordinary feats of uncanny foresight, beginning in 2017, Dr. Fauci paid $6 million to his gain-of-function guru, Ralph Baric—a University of North Carolina microbiologist—to accelerate remdesivir as a coronavirus remedy. . . . Baric used coronavirus cultures obtained from bat caves by Chinese virologists working with Peter Daszak's EcoHealth Alliance, another recipient of Dr. Fauci's funding. . . . Baric claimed that his mouse studies showed remdesivir impeded SARS replication, suggesting that it might inhibit other coronaviruses.[106]

Kennedy's book—a magisterial work of scholarship for which he vetted thousands of cited sources—details all the strange twists in the process of obtaining an EUA for remdesivir. The most salient was Dr. Fauci's stunning decision to change the trial's endpoint after the drug failed to meet the original—namely, the reduction of COVID-19 mortality:

> Dr. Fauci's new endpoints allowed the drug to demonstrate a benefit, not by improving the chances of surviving COVID, but by achieving shorter hospital stays. Yet this too was a scam, because it turned out that almost twice as many remdesivir subjects as placebo subjects had to be readmitted to the hospital after discharge—suggesting that Fauci's improved time to recovery was due, at least in part, to discharging remdesivir patients prematurely. Altering protocols in the middle of an ongoing study is an interference commonly known as "scientific fraud" or "falsification."[107]

Per standard clinical trial practice, changing the endpoints after a trial has started is to invalidate the trial. And as Kennedy pointed out, it was doubtful the revised endpoint was met. Doctors conducting the study knew from telltale high liver function tests which patients were receiving the remdesivir, which meant their study wasn't blinded. They were therefore potentially biased in favor of discharging these patients earlier than the control patients.

Before the NIAID's remdesivir trial was peer-reviewed and published, the *Lancet* published a Chinese study showing remdesivir produced no benefit in reducing mortality or hospital duration or clearing the virus.[108] Even worse, a second Chinese trial was halted after 12 percent of the experimental group experienced severe side effects including liver and kidney damage.[109]

Nevertheless, at the April 28 White House press briefing, Dr. Fauci announced, "The data shows that remdesivir has a clear-cut, significant, positive effect in diminishing the time to recovery." He elaborated that the remdesivir study group was able to be discharged from the hospital within eleven days, on average, compared to fifteen days in the placebo group.

"What it has proven is that a drug can block this virus," Dr. Fauci said. The trial results were "reminiscent of thirty-four years ago in 1986 when we were struggling for drugs for HIV."[110]

Three days later, the FDA granted remdesivir an EUA, and a month later President Trump purchased the entire existing stock for Americans.[111] Thus we see how one man, Dr. Fauci, guided remdesivir—an experimental drug with a questionable history, doubtful efficacy, and grave safety concerns—to become the standard of care in the United States and a bonanza for Gilead Sciences.

The story doesn't end there. On October 8, 2020, the European Commission signed a contract with Gilead for 500,000 treatment courses of remdesivir at EUR 2,070 per course for a total cost of EUR 1.035 billion.[112] According to a query submitted to the European Parliament by French MP Virginie Joron:

> On 9 October 2020, the European Union was informed of the negative results of a study [Solidarity trial] carried out by the World Health Organization (WHO) in 405 hospitals in 30 countries covering more than 11,000 COVID-19 patients. That study, which was published on 15 October 2020, absolutely ruled out "the suggestion that remdesivir can prevent a substantial fraction of all deaths." On 20 November 2020, the WHO officially advised "against the use of remdesivir in hospitalized patients, regardless of disease severity, as there is currently no evidence that remdesivir improves survival and other outcomes in these patients."[113]

This wasn't the first time the European Union was left holding the bag of expensive pharmaceutical products of doubtful value. The Commission made similar emergency purchases of dubious products during the Swine Flu Pandemic of 2009, which proved to be a false alarm:[114]

> Even after the WHO published the negative results of its Solidarity trial on October 15, 2020[115] and recommended *not* using remdesivir, the NIH continued recommending the drug for hospitalized patients. On October 20, 2020, the FDA fully approved it under the trade name Veklury for use in adult and pediatric patients 12 years of age and older.

CHAPTER 14

"According to my ability and judgment"

Around the same time that Dr. Fauci recommended remdesivir, Dr. McCullough's clinical trial of hydroxychloroquine prophylaxis for hospital staff was moving along. Apart from a few isolated cases of mild nausea, no side effects were reported in the experimental group. Nevertheless, as the media campaign against hydroxychloroquine reached fever pitch, someone complained to the hospital's Institutional Review Board that McCullough was conducting "a dangerous drug experiment on staff members." This initiated an administrative review of the study and the imposition of new requirements that made it considerably more difficult and time-consuming to carry out.

The media campaign against hydroxychloroquine had the effect of maligning *all* forms of early treatment. This was terribly unfortunate, because as McCullough was learning, there were other facets of the disease that could be addressed with other therapies. The hospital's nephrologists noticed blood clotting in the disease's advanced stage. Some patients were placed on dialysis to treat kidney failure, and the machine's lines were clotting up. This was a significant observation, and it raised the possibility of intervening with anticoagulants. Around the same time, McCullough saw credible reports on the benefit of corticosteroids to reduce inflammation.

One of the older nurses in his research study who'd chosen to forgo the hydroxychloroquine got fairly sick, and her symptoms concerned him. She was scared and desperately wanted to avoid being hospitalized. He felt duty-bound to treat her as best as he could, so he prescribed to her a daily regimen of hydroxychloroquine, azithromycin, the anticoagulant Eliquis, and the steroid prednisone. She seemed to respond favorably to the treatment, but just a couple of days into it, something strange happened. The hospital's chief medical officer contacted him.

"Dr. McCullough, I heard you're using a bunch of drugs to treat a nurse for COVID-19 as an outpatient."

"Yes, I'm a bit worried about her and I want to keep her out of the hospital."

"Do you realize you have no scientific support for doing this?" she said in an accusatory tone.

"Well, we're learning about this, and I don't feel comfortable doing nothing, so I'm going to treat her according to my clinical judgment," he said. The officer backed off and the patient recovered. Nevertheless, it was an extremely jarring incident. How did the medical officer even know the specifics of McCullough's treatment of the nurse? This was a breach of doctor-patient confidentiality. Chief medical officers, whose job was to enforce the hospital's physician discipline policy and adjudicate other activities on hospital operations, were *not* in the business of calling senior doctors to question their judgment or treatments and tell them what generic drugs they could prescribe for any condition.

The medical officer's posture had also seemed strangely unsympathetic, perhaps even hostile to treating the nurse with safe and commonly used drugs. Nor did she seem to have any regard for the nurse's desire to try the therapy. For McCullough, it wasn't only his sense of duty to his patient, his adherence to the Hippocratic Oath. He also knew the nurse and admired her work ethic. Nurses, more than members of any other profession, were essential for dealing with the pandemic.

McCullough's second experience treating COVID-19 happened around the same time. One of his patients was suffering from heart and lung disease and breast cancer. Upon returning to Dallas from MD Anderson in Houston, she developed a fever and malaise and took a hard fall in her backyard. At a hospital near her home, staff took X-rays and drew some blood for analysis. Strangely enough, her lymphocyte count was depressed, and it couldn't be explained by her cancer treatment. McCullough had heard talk of this finding in COVID-19 patients. Normally with a viral infection, the lymphocyte count goes up. The only exception that McCullough knew of was HIV. He suspected the patient had COVID-19—though no test was conducted at the time—and he put her on hydroxychloroquine +

azithromycin + lovenox. With her three comorbidities, she got quite sick but ultimately pulled through. Later she tested positive for COVID-19 antibodies. And so, McCullough was learning to recognize and fight the disease. His newly acquired knowledge was about to be put to the test.

The trouble had begun in January, when his 83-year-old father, Thomas McCullough, fell and broke his pelvis. Thomas and his wife, Mary, were living in Fairhope, Alabama, where they'd moved for their retirement fifteen years earlier. Thomas's grave injury spelled the end of it. For a while he resided in the Robertsdale Rehab Center, about twenty miles from their home, but his recovery was agonizingly slow, and forty miles of daily driving was a hardship for Mary. At the end of February, McCullough traveled to Alabama to assess the situation and saw it was untenable.

"I'm afraid you and Dad are going to have to move back to Dallas where I can get better care for you," he told his mother. She consented, and he organized a luxury ambulance to drive his father twelve hours straight from Alabama to the Presbyterian Village nursing home and rehab center in Dallas. Thomas arrived in remarkably good spirits, but scarcely recognizable with his long, ill-kempt beard, indicating the rehab center had suspended his hygienic care while he was stuck on his back. At first glance McCullough thought he looked like a homeless person, which he found upsetting.

Things went well during the first weeks of March. At the end of each workday, McCullough and his wife, Maha, and their dog went to Presbyterian Village to visit his father. During this time, his father took his first, agonizing baby steps and reached the significant milestone of being able to use the toilet. McCullough felt a glimmer of hope that his dad might well enjoy a full recovery. The old man seemed to draw strength and encouragement from his son's daily visits. Then the restrictions started.

One day they arrived at the entrance guardhouse to check in and were questioned about their health. A couple of days later their body temperatures were checked. Then, in another day or two, they arrived to find parking cones across the entrance. No visitors were allowed. Soon McCullough's mother began to receive calls from his father. Suffering from dementia, he was confused about his isolation and didn't understand why no one was visiting him. For the first time in his adult life, he was unable to see his wife. They had been together through thick and thin for a long time, and he was profoundly disoriented by her absence. Often in the middle of the night, McCullough was awakened by his father calling and asking, "Where am I?" and "How do I get out of here?"

Thomas McCullough was the son of modest Scotch-Irish immigrants who wound up in a Hell's Kitchen tenement building with poor sanitation, later to resettle in New Jersey, then Buffalo, NY. As a child, his father had contracted rheumatic fever, which damaged his heart's mitral valve. When he was 52, he underwent a primitive open-heart operation in which the surgeon attempted to invaginate the constricted valve by pressing on the exterior of the left atrium with his thumb. The procedure, known as a closed commissurotomy, didn't work, and his father died shortly thereafter, leaving Thomas to essentially inherit his position at the Western Electric Company's manufacturing plant in Buffalo. A wave of layoffs in the early seventies left him out of work with a wife and three boys and little money.

Somewhere he read that the Sprague Electric Company in Wichita Falls, Texas, was hiring and that the small city 140 miles northeast of Dallas had affordable housing. And so he loaded his family Ford LTD station wagon with their most valuable possessions, and—with McCullough and his two brothers riding in the backseat and their mom riding shotgun—they drove from Buffalo to Texas. Decades later, McCullough still remembered crossing the Red River, and then pulling into the Trade Winds Motel in Wichita Falls. The August day was hotter than hell and windy, and everything was covered with red dust. For a week the family lived in a single motel room—like their Irish ancestors in the New York City tenements of old—while his father got situated with a job and bought a little house on the edge of a field.

It wasn't an auspicious start in Texas, and McCullough thought his father had seemed defeated. But then he'd shown true grit and landed a good job at the Abbott Laboratory facility in Irving, Texas, just west of Dallas. McCullough would always remember that day they pulled into Wichita Falls, thereby changing his destiny. Had his family remained in Buffalo, he would have likely followed in his father's footsteps as a factory worker. In Texas his studious habits opened a new path leading to a Baylor University scholarship.

There are times in every son's life when his father seems less than admirable. McCullough had sometimes lamented his dad's wild Irish ways—his rough manners and language and tendency to drink too much. But seeing him so vulnerable, McCullough thought about how the old man had always persevered for his family as best as he could. Now his late-night pleas for help broke McCullough's heart.

When is this nightmare going to end? he wondered. President Trump spoke of reaching the other side by Easter, which fell on April 12. That day arrived with warm sunny weather in Dallas but no end in sight. A couple of days later McCullough was informed that a staff member at Presbyterian Village had come down with COVID-19, but that she'd had no contact with his father. Then his mother called to say that his dad's spring allergies were flaring up, as he sounded congested on the phone. Then he got the call that his dad had tested positive for the virus.

By then McCullough knew that isolating the disoriented man from his wife hadn't protected him because he was still exposed to nursing home staff. While his wife had little contact with anyone, staff members had close contact with innumerable persons in multiple facilities every day. So now his father—an octogenarian suffering from a multifractured pelvis and senile dementia—had symptomatic COVID-19.

McCullough's mother informed the facility's administrators that her husband had long ago signed a directive stating that he never wanted to be kept alive by artificial life support. This meant that, in the event of acute respiratory distress, sending him to the hospital to be put on a ventilator was out of the question. Considering this, the facility made way for a COVID unit on the top floor of its Corrigan Administrative building, and Thomas McCullough was their first patient.

How to save his father's life? Disabled by the pelvic fracture, he spent most of the time on his back, which reduced his respiratory mechanical function. Judging by the mortality data out of China and Milan for a patient his age with his health problems (including diabetes), his chance of survival was probably around 20 percent. McCullough knew that his newly acquired knowledge was fragmentary and fraught with uncertainty. *According to my ability and judgment.* He would just have to do the best he could.

By a stroke of luck, a doctor's assistant at the facility was open to suggestions from McCullough, who was immensely grateful for her and the nurses who looked after his father.

"The weather is beautiful, so the first thing we do is open all the windows to get fresh air into the room to reduce the inoculum," he told her. The doctor's assistant was game to do this.

"Secondly, let's put him on a twice-daily dose of hydroxychloroquine and azithromycin, plus aspirin and lovenox."

In the days that followed, McCullough's father got very sick and lost his appetite. Dehydration set in, and his blood pressure dropped. They tried to administer IV fluids, but his veins had retracted too much to place the needle. And so, McCullough instructed the staff to perform the pediatric method of sticking the needle into the abdominal fat. They later succeeded in drawing a vial of blood and saw that his serum sodium level was 151 milliequivalents per liter, which meant he was catastrophically dehydrated. Finally, they succeeded in getting enough fluids in him and kept him on the medications for thirty days. He lost forty-two pounds, but he survived. Through the illness, the nurses noted that all he wanted was protein shakes. This turned out to be a wise choice for a diabetic, for as would soon be discovered, severe disease strongly correlated with high blood sugar.

There's an archetypal myth at least as old as the ancient Egyptian god Horus, who retrieved his father from the underworld. For the child, the father represents order and security. The passage from childhood innocence to full adult awareness is expressed as venturing into an extremely dark place to rescue one's father. This involves being shorn of all comforting illusions and security and facing the full horror of our mortal existence. At the age of 57, McCullough had seen and learned a lot. He'd completed vast medical training, gotten married, fathered two children, and held leadership positions in major hospitals.

And yet, he'd always lived in a structured environment governed by well-established procedures. Most aspects of his life had seemed settled on the assumption he would finish his career at Baylor and enjoy a comfortable retirement. Within this powerful institution, he held a ranking position that gave him high social status and generous remuneration and health benefits. His professorship at its affiliated medical school provided him with additional authority and prestige. As most of his colleagues could attest, reaching such a position was the culmination of decades of study and work.

Undertaking the task of treating his father for the novel disease required going far beyond the familiar routines and procedures of his institution. The anguish of seeing his old dad laid low, and the victory of bringing him back from death's threshold, was a trial from which he emerged with an expanded sense of himself and his responsibility. It prepared him to play a role in human affairs far beyond that of a conventional doctor.

Just as he could not and would not sit back and do nothing while the virus killed his father, he wasn't about to sit back and do nothing for his patients. Following his conscience, he knew he had to advocate treating COVID-19 for all of mankind. For contrary to what government agencies and the media were saying, it *is* a treatable illness.

CHAPTER 15

On the Front Line of Critical Care

Thirteen hundred miles to the east, in Norfolk, Virginia, another doctor was embarking on a similar path. Dr. Paul Marik was to critical care medicine what McCullough was to cardiorenal medicine. The 64-year-old native of South Africa had published over 500 peer-reviewed papers and books, which made him the second-most published critical care doctor in the history of medicine.

Upon meeting Dr. Marik, one is overwhelmed by the impression of his vast physical stature. His gentle manners and elegant South African accent conjure accounts of 19th-century British gentleman explorers. Since the early days of his career, he'd been keenly interested in discovering how to treat sepsis—the body's extreme, life-threatening response to an infection. Well into the 21st century, sepsis continued to be a major cause of death. According to the CDC, approximately 1.7 million American adults fall ill with sepsis every year, of which approximately 270,000 die in hospital.[116] Globally, the sepsis burden is estimated at 15 to 19 million cases annually, with a mortality rate approaching 60 percent in low-income countries.

Dr. Marik knew the literature on sepsis treatment. Several studies had indicated that large doses of IV-administered Vitamin C and Vitamin B1 (thiamine) showed some benefit. Other studies indicated that hydrocortisone showed promise. Dr. Marik reasoned that combining the three into a cocktail could amplify their benefit. As had often been observed in medicine, combining agents seems to affect multiple pathways, causing an overlapping and synergistic effect.

In 2016, he and his colleagues conducted a study in which they compared forty-seven sepsis patients who received the cocktail to a control group of forty-seven patients who received sepsis standard of care at the time. The mortality rate of the treatment group was 8.5 percent compared to 40.4 percent of the control group—a stunning difference.[117] Skeptics claimed the study was too small and nonrandomized. Nevertheless, what became known as the "Marik Cocktail" was adopted by critical care units all over the world, which reported excellent results with the therapy.

When COVID-19 struck and Dr. Marik's critical care unit at the Sentara Norfolk General Hospital received its first patients, he observed they were suffering from an inflammatory lung reaction, and he suspected this could be treated with a corticosteroid. He contacted four other critical care specialists—Professor Joseph Varon at United Memorial Medical Center in Houston (who, in the year 2020, would work 268 days straight treating COVID-19 patients in his ICU); Professor Gianfranco Umberto Meduri at the University of Tennessee Health Science Center in Memphis; Professor Jose Iglesias at the Hackensack Meridian School of Medicine in Seton Hall, New Jersey; and Professor Pierre Kory at the University of Wisconsin-Madison. It would be hard to imagine a more experienced and scholarly team of pulmonary critical care doctors.[118]

Together they formed the Front Line COVID-19 Critical Care Consortium (FLCCC) and got to work on a protocol for saving hospitalized patients. They started their work by focusing on the extreme inflammation they were observing. Often called "cytokine storm" after the proteins produced by the immune system, this was an extreme and maladaptive immune response that had often been observed in other viral illnesses, including virulent influenza. It was this inflammation of the lungs, kidneys, and other organs that killed the patient, and not the virus itself. Thus, the key to treating the syndrome was reducing this inflammation.

Professor Meduri was a leading expert on corticosteroid therapy in critical illness. He conducted an exhaustive review of corticosteroid use against SARS-CoV-1, MERS, and H1N1 and found significant evidence that these agents had been lifesaving in their reduction of extreme inflammation. For decades the corticosteroid methylprednisolone had been the standard medication for suppressing cytokine storm. And yet, despite this glaring fact, every US federal health agency recommended *against* using corticosteroids against COVID-19 from the outset of the pandemic.[119]

Unlike these agencies, the FLCCC concluded that their best bet for tackling the disease in hospitalized patients was a combination of

Methylprednisolone and the antioxidant **A**scorbic acid (Vitamin C). To these they added **T**hiamine (Vitamin B1) to optimize cellular oxygen utilization and energy consumption and the anticoagulant **H**eparin to prevent and help in dissolving blood clots that appear with high frequency.

There was nothing experimental about these drugs. Methylprednisolone and heparin had long been FDA-approved for treating inflammation and blood clotting. Vitamin C and B1 were available over the counter, though in the hospital setting they were IV-administered in high doses. The FLCCC doctors began giving this "MATH Protocol" to ICU patients and tracked their progress for a case study. By the end of April, they had treated 100. Of these, 98 survived. The two who died were in their eighties and had other advanced chronic illnesses. None of the patients had long durations on ventilators, were ventilator dependent, and/or had long hospital stays.

During this period, Dr. Pierre Kory was in daily communication with ICU doctors in his native New York City, where he'd worked in multiple hospitals. His colleagues reported that they turned the tide in the ICUs as soon as they started using methylprednisolone. Other ICU doctors in hard-hit New Orleans reported the same. Kory and colleagues therefore sent letters to the White House, the CDC, and the NIH, presenting their real-world evidence of the corticosteroid's efficacy—all to no avail.

Dr. Kory's efforts were drawn to the attention of Wisconsin Senator Ron Johnson, who was chairman of the Senate Homeland Security Committee. Since his daughter had been born with a congenital heart defect—successfully corrected by an innovative operation—Senator Johnson had believed that if citizens were faced with a life-threatening condition for which there was no proven treatment, they should have the right to try drugs or operations that seem to offer some benefit, even if these had not yet passed the conventional FDA-approval process. The alternative was to do nothing and accept the high probability of death.

Critics of the right-to-try claim that unless procedures and medications are carefully regulated by the FDA, unscrupulous doctors will create false hope by offering them to desperate patients, even if there is little trial data to prove their efficacy. Senator Johnson did not find this argument persuasive, and he introduced his federal right-to-try bill in 2017. A companion bill was introduced in the House, ultimately passed in both houses, and was signed into law by President Trump in 2018.

Upon Senator Johnson's invitation, Professor Kory addressed the Senate Homeland Security Committee on May 6, 2020. Speaking via WebEx, Kory stated the credentials of the FLCCC doctors and then explained the rationale for using their protocol. He then reported the success they and other doctors were having with it.[120] This was excellent news, and a naive viewer would likely assume that it would be welcomed by the entire medical profession.

And yet, despite the FLCCC's well-founded rationale for their therapy, broad consensus for the efficacy of its components in related conditions, and their success with it, they continued to get pushback from the CDC and NIH, which refused to change their advisory against using corticosteroids to treat COVID-19. On four occasions, the FLCCC tried to notify the White House of their favorable results, but so far, they'd received no reply.

CHAPTER 16

"What would Gene Roberts have done?"

Among those who saw a recording of Dr. Kory's testimony was the journalist and bestselling author Michael Capuzzo. The four-time Pulitzer Prize nominee had a keen sense of people and their credibility, and he got the overwhelming impression that Dr. Kory was speaking the truth. The doctor wasn't guessing or speculating or indulging in wishful thinking. He and his colleagues knew what they were talking about.

Capuzzo knew from history how long-standing assumptions may, at any time, be challenged by unexpected events. Early in his career he'd written a feature on Dr. Sheila Katz, a pathologist at Philadelphia's Hahnemann Medical College and Hospital. Dr. Katz had been on the front line of studying a mysterious outbreak of a lethal pneumonia at the Bellevue Stratford hotel in Philadelphia. On July 21, 1976, the Pennsylvania section of the American Legion and their families arrived at the grand hotel to celebrate their annual jamboree. In the days following the event, 151 Legionnaires fell ill with pneumonia, and 29 died. All the victims had just been at the Bellevue Stratford.

The CDC sent its best and brightest to the hotel to figure out the cause. The lead investigator, David Fraser, was a 32-year-old Harvard Medical School graduate who bore an uncanny resemblance to Bobby Kennedy. In spite of his extraordinarily thorough investigation, he was unable to determine what pathogen had caused the outbreak. He and his senior colleagues at the CDC found it perfectly baffling. As Director David J. Sencer testified on November 24, 1975: "The outbreak ... has presented a number of unusual and complex features. . . . It has run counter to our expectations that contemporary science is infallible and can solve all the problems that we confront."[121]

Dr. Katz herself fell ill with severe pneumonia after studying a lung specimen from one of the legionnaires who'd died of the disease. She survived the ordeal, which deepened her interest in the mystery. Ultimately it was a 36-year-old "backbencher" CDC microbiologist named Joseph McDade who solved it. While his colleagues were searching for pathogens they already knew about, Dr. McDade pursued what appeared to be an unlikely lead—a single rod-shaped bacterium that he caught a vague glimpse of in his microscope. His more experienced colleagues regarded it as a random, solitary bacillus of no consequence, which initially prompted McDade to dismiss his own observation.

Months later he returned to the specimen and discovered it was a hitherto unknown Gram-negative bacterium that was difficult to see using standard Gram-negative stain. The bacterium, which was given the name *Legionella pneumophila,* can contaminate air-conditioning cooling towers such as the roof unit on the Bellevue Stratford. It took the CDC six months to solve the mystery of Legionnaires' disease. At a moment in history when scientists thought they'd seen it all, they were humbled by nature. The moral of the story was clear: *Don't only look for things that you already know about, but keep your mind open to the possibility of new discoveries.*

Capuzzo knew from this and many other stories he'd researched that scientific understanding is not fixed, but constantly evolves through new experience and observation. In researching his 2001 bestseller, *Close to Shore*, about the summer of 1916 Jersey Shore shark attacks (the inspiration for the novel and film *Jaws*), he delved deep into the literature on sharks and learned how little was known about the creatures, even in the 1970s, when Peter Benchley wrote his sensational novel about a rogue Great White terrorizing Amity Island.

And so, when Capuzzo saw the video of Dr. Kory's May 9 Senate testimony, he sensed it was a story for the ages. He set about researching Dr. Marik's FLCCC team. They were, as he would later write, "a group of intensivists with nearly 2,000 peer-reviewed papers and books and had over a century of bedside experience in treating multi-organ failure and severe pneumonia-type diseases. If anyone could arrest the coronavirus in a living patient, they could."[122] Why weren't federal health agencies listening to them?

Even more bizarre to Capuzzo was the fact that none of the nation's top newspapers were reporting the story. He knew the newspaper world from his six years as a reporter for the *Miami Herald* and eight for the *Philadelphia Inquirer*. He'd also written pieces for the *Wall Street Journal*, *New York Times*, and *Washington Post,* and he still had close friends among their editors and staff writers. He contacted them and pitched a story on Dr. Marik and the FLCCC. None showed interest. He got a nibble at the *Wall Street Journal* and wrote a report on Dr. Marik's undeniable success in saving critically ill patients with his MATH protocol—patients with the highest risk profile, brought back from the dead. For some reason, the *Journal* chose not to run it.

Often, he asked himself: "What would Gene Roberts have done if I came to him with the FLCCC story?" Reflecting on his former *Philadelphia Inquirer* editor, he wrote in an e-mail:

> Eugene Roberts was the legendary editor who left the *New York Times*, where he was the national editor, to take over the moribund *Inquirer* in 1972, and by 1976 was well on his way to leading the *Inquirer* to 17 Pulitzer Prizes in 18 years as one of the best papers in the country, with Mark Bowden, Buzz Bissinger, etc. on staff.
>
> What would Gene Roberts have done? He would have his reporter or two or three, with a photographer, go to Dr. Marik's hospital and talk to people who were living and not dying. Science and medical writers would profile Dr. Marik and call around vetting his research and include statements from skeptics at the CDC. He wouldn't let the NIH or CDC kill the story because "the scientific evidence" wasn't yet in. On the contrary, the story would be ON THE FRONT PAGE.[123]

Clearly things had changed since those lost days of 20th-century journalism.

Capuzzo's interest in the story deepened as Drs. Marik and Kory learned about the extraordinary value of ivermectin in treating COVID-19. He did a deep dive into the literature on the wonder drug and was again astonished that no major newspaper was reporting its promise. Fairly soon it became apparent to Capuzzo that he probably wasn't going to tell the story by way of the legacy media with which he'd always worked. Among veteran American newspaper reporters, he was apparently alone.

CHAPTER 17

Nihilism and Fraud

Back in Dallas, Dr. Peter McCullough continued to work on developing early, outpatient treatment. Though he applauded the efforts of Drs. Marik and Kory to care for their patients in the hospital, he believed the primary focus should be to prevent patients from going to the hospital in the first place. And yet, over two months into the pandemic, the citizenry still wasn't receiving any constructive advice, hope, or even information. None of the major university medical schools such as Harvard or Johns Hopkins attempted to develop a treatment protocol. Conferences to discuss possible therapies were virtually nonexistent, and there were no publicized lines of communication for reporting real-world observations from doctors treating patients in the field. Repurposed drugs such as hydroxychloroquine were immediately shot down by federal health and academic medical authorities.

Therapeutic nihilism, McCullough thought. There was no other way to describe it. He often wondered: *Since when did the medical profession categorically give up on the idea of treating a disease?* It's a fundamental principle of medicine that if a disease is known to progress in severity, possibly to hospitalization and death, it's imperative to intervene as early as possible to try to stop it. And yet, in the case of COVID-19, patients diagnosed with the illness were told to go home and wait to see if it progressed to shortness of breath, at which point they should go to hospital.

Not only was the biggest healthcare system in the world doing *nothing* to help sick Americans; its leaders couldn't even define the problem. The media's relentless reporting of raw case and death counts—tallied by the Johns Hopkins Medicine IT department—conditioned the public to focus obsessively on them while inextricably associating "cases" with "deaths." McCullough wondered about the accuracy of these reports. How was the cause of death being ascertained and reported so quickly? Normally the issuance of a death certificate took several weeks.

What was being reported as "cases" were positive PCR tests for the presence of the SARS-CoV-2 virus, or fragments of the virus, found in nasopharyngeal swabs. Following basic principles of infectiology, this positive test alone, with no accompanying symptoms, is insufficient to diagnose the disease. The inventor of the PCR test, Kary Mullis, once stated, "it can find almost anything in almost anyone," and he warned against it being used to determine an infection.[124]

Positive test data were also being inflated by duplicate tests and false positives from improperly elevated PCR test cycle thresholds. Large CARES Act payments to hospitals for admitting COVID-19 patients created a perverse incentive to enter this code even for patients suffering from other illnesses and injuries. If you tested positive for SARS-CoV-2 and died of an unrelated myocardial infarction, your death could be registered as a COVID-19 fatality. Overstated positive test data followed on the heels of wildly inflated theoretical projections of COVID-19 hospitalizations and deaths from computer modeling. All of these distortions amplified fear and distrust and seemed to justify investing public officials with emergency authority to impose lockdowns and other social controls.

Yet another distortion was the constant talk in the media of "asymptomatic spread." This enabled test manufacturers and centers to make fortunes, but there was no scientific basis for believing it. Among the symptomatic, those under fifty with no comorbidities faced little risk of developing severe COVID-19, so it was misleading to present these "cases" as though they were the problem.

The real problem consisted of hospitalizations and deaths that were verifiably caused by COVID-19. The solution to this problem was to identify and *treat* those at risk of severe illness in order to prevent them from going to hospital. McCullough marveled that this obvious proposition wasn't even being articulated, much less acted upon.

Accurate hospitalization data were the key to working up a profile of who was getting gravely ill from the virus, which was the only way to determine the severity of the public health threat. Obvious risk factors for severe disease—advanced age, obesity, and diabetes—had been identified, but medical science is in the details. And so, McCullough published an op-ed in *The Hill,* an independent political news site/journal in Washington, DC, calling upon the Trump Administration to order the daily publication of hospital census data. He urged the same action in a peer-reviewed-paper in *Reviews in Cardiovascular Medicine.* Both pleas apparently fell on deaf ears.

It was as though the entire academic medical system had totally surrendered to SARS-CoV-2. This was especially strange given how much was already known about the virus. It was already fully sequenced and rapidly detectable by standard testing. Its target receptor (ACE2) and dangerous spike protein were understood. The disease's distinct stages and pathologies generated at each stage were also known. The most significant—cytokine inflammation and blood clotting—could be treated with drugs that had long been widely used.

All the above was understood. Now it was time for someone with McCullough's credentials to publish an academic paper describing the "pathophysiological basis" for treating the disease. Once this paper was in the literature, he and his colleagues could cite it for their practices.

His starting point was the protocol he'd already used successfully on a handful of patients—hydroxychloroquine + azithromycin + zinc to inhibit viral replication, steroids to inhibit inflammation, aspirin and lovenox to inhibit blood clotting. McCullough thought about specialists who could provide coauthoring expertise. Just as he was getting organized, the world of academic medical publishing was rocked by a bizarre scandal.

For almost two centuries, doctors all over the world had shared knowledge through medical journals. The dissemination of new discoveries helped doctors everywhere to provide better care for their patients. In recent years, online electronic publishing of papers facilitated the sharing of information.

Since 1823, the *Lancet* has been one of the most prestigious medical journals in the world. It had maintained its reputation through a careful vetting of submissions. This process starts with an examination of the authors, their backgrounds, and their publishing records. Their submission—its data, methods, and conclusions—is then circulated to peers for review, and then the editors perform a close reading.

On May 22, the journal published a metastudy by Dr. Mandeep R. Mehra of the Harvard Medical School and Dr. Sapan S. Desai of the Surgisphere Corporation. The study presented an analysis of the electronic records of nearly 100,000 COVID-19 patients in 671 hospitals on 6 continents who were given hydroxychloroquine to treat COVID-19. The authors concluded that the drug increased the risk of death in patients who received it compared with controls.[125]

The consequences of this paper were swift and dramatic. As a review of the incident in *The Scientist* put it:

> The study was a medical and political bombshell. News outlets analyzed the implications for what they referred to as the "drug touted by Trump." Within days, public health bodies, including the World Health Organization (WHO) and the UK Medicines and Healthcare products Regulatory Agency (MHRA) instructed organizers of clinical trials of hydroxychloroquine as a COVID-19 treatment or prophylaxis to suspend recruitment, while the French government reversed an earlier decree allowing the drug to be prescribed to patients hospitalized with the virus.[126]

When McCullough read the paper, he instantly recognized that its dataset was implausible. The number of hospitalized patients and the death count were impossibly high for the study's period, at the very beginning of the pandemic. Likewise, the average age of the patients was forty-nine. He knew there was no way that so many relatively young people had been hospitalized from COVID-19.

And what on Earth was Surgisphere? A bit of digging revealed it had been founded in 2008 by Dr. Desai while he was a surgical resident at Duke University. Since then, questions about his integrity had arisen. In February of 2020, he left his position at Northwest Community Hospital in Chicago while being sued for three incidents of medical malpractice—two involving permanent damage following surgery and one involving a patient's death. Around this same time, he launched a new Surgisphere website, replete with articles asserting collaborations that never took place.[127]

McCullough wasn't the only reader who questioned the paper. Epidemiologists and statisticians sent letters to the *Lancet*'s editor, Richard Horton, expressing concern about its data and conclusions. On June 5, the *Lancet* retracted the paper after independent auditors were denied access to Surgisphere's full dataset. As was posted on its website:

> Based on this development, we can no longer vouch for the veracity of the primary data sources. Due to this unfortunate development, the authors request that the paper be retracted.[128]

While the paper's publication had immediately resulted in the suspension of multiple hydroxychloroquine trials, there were significant delays in resuming them after the paper was retracted. Likewise, while the mainstream media had pounced on the paper's publication, the MSM ignored its retraction. Thus, to a large extent, the damage to hydroxychloroquine research—and to perceptions of the drug within the medical profession and general public—was irrevocable.

Judging by what happened in Switzerland, the fraudulent *Lancet* paper had deadly consequences. As one prominent observer noted:

> On May 27, the Swiss national government banned outpatient use of hydroxychloroquine for COVID-19. Around June 10, COVID-19 deaths increased four-fold and remained elevated. On June 11, the Swiss government revoked the ban, and on June 23 the death rate reverted to what it had been beforehand. People who die from COVID-19 live about three to five weeks from the start of symptoms, which makes the evidence of a causal relation in these experiments strong.[129]

McCullough wondered how such an obviously fraudulent paper had passed through peer review and editing. He found everything about the incident unfathomably strange. *Holy smokes, this is getting alarming,* he thought. *The high-level journals were being corrupted before our very eyes.*

Thankfully, five days after the fraudulent *Lancet* paper appeared, a paper was published in the *American Journal of Epidemiology* that was a pretty good counterpunch. The author was a professor at the Yale School of Public Health.[130]

CHAPTER 18

Professor Risch Punches Back

Like Richard Feynman, with whom he'd studied physics at Cal Tech, Harvey Risch had a wide-ranging intellect and a long-standing habit of questioning scientific orthodoxy. Feyman taught him that scientific discovery often began with exploration unconstrained by the prevailing assumptions of the day. After graduating from Cal Tech, Risch went to medical school to earn his MD, where he discovered a love of research. After medical school he earned a PhD in biomathematics and wrote his dissertation on mathematical modeling of infectious disease outbreaks.

Upon meeting him, one is immediately impressed by his calm, fastidious, and disciplined demeanor. A practicing Jew who wears a kippah, he comes off as the embodiment of scholarship, both in the metaphysical and physical realms. For much of his academic career he focused on cancer epidemiology and published 350-peer-reviewed papers on the subject. Shortly after COVID struck, the Connecticut Academy of Science and Engineering invited him to join an expert panel to formulate a plan for dealing with the disease. He read the literature on which treatments had shown promise so far. Reports from doctors in the field who were treating patients with hydroxychloroquine combined with azithromycin and zinc sounded promising. And yet, to Professor Risch's puzzlement, they were quickly dismissed by public health agencies and university hospitals. Why?

By April, the most common rationale was that studies conducted on hospitalized patients had purportedly found no benefit. The conclusion was "the drug doesn't work in hospital, so it therefore doesn't work at all." But this made no sense to Professor Risch. As he'd long understood, science is in the details. Doctors in the field reported that the early treatment of COVID with hydroxychloroquine was preventing their patients from going to hospital in the first place, so what was the sense in drawing conclusions about the drug's efficacy based on its use in hospitalized patients? By then the viral illness had advanced to a later stage with a host of pulmonary and systemic problems, at which point any medical intervention would have far less chance of success. Like Professor Didier Raoult in France, Professor Risch wondered if lab scientists were so myopically focused on questions of method that they were unable to conceptualize the problem at hand.

Since prehistoric times, humans have learned about the natural world by observing patterns and associations. An association observed a few times suggests that a certain action causes a certain effect. Even if people don't know for sure if there is a causal link, repeated observations are often enough to guide effective action. In the realm of medicine, humans used plant-based remedies such as willow bark (salicylic acid) and cinchona bark (quinine) for centuries. Their use was guided by the observation that pain diminished in those who took the former and malaria didn't sicken those who took the latter. Though humans had used willow bark, from which aspirin was derived, for centuries, aspirin's mechanism of action wasn't discovered until 1971.[131]

Since the Scientific Revolution, man has developed experimental methods for distinguishing random associations from the causally linked. A principle method used by the pharmaceutical industry is the Randomized Controlled Trial (RCT). As the industry has grown in size and sophistication, it has increasingly asserted that *only* RCTs yield valid evidence. However, as Professor Risch had learned from years of study, this is not true.

Scientific discovery often begins with observations in the field and lab. In 1928, Alexander Fleming was looking at a bacteria culture on an agar plate in his laboratory at St. Mary's Hospital in London and saw that a mold was also growing on the plate. Taking a closer look, he observed a zone around the mold in which no bacteria would grow. Quickly he deduced that the mold possessed an antibacterial property. He then extracted the active agent and named it penicillin. This random observation was the origin of what was probably the most useful drug ever developed—a drug whose efficacy was never tested with an RCT.

Many of the great techniques in trauma treatment were made by military doctors in combat theaters. Likewise, the treatment of eye injuries made a great leap forward in the medical response to the Halifax explosion of 1917, when a munitions ship detonated in the harbor and blew glass and other debris into the eyes of hundreds. The doctors from Boston who treated them had no time to perform experiments. They did the best they could and documented their observations.[132]

A treating physician administers a drug to sick patients and observes that within a day or two they feel better instead of worse. He then applies this observation to a case study of *high-risk* patients and observes that a much smaller percentage of them go hospital than untreated *high-risk* patients from the general public. Such a case study is not an RCT, but it's still valid evidence, and very valuable for treating sick people.

In 1847, when Professor Semmelweis asked his medical students to wash their hands after working with cadavers and before they examined pregnant women, he was using elementary deductive reasoning. The benefit he observed—a reduction of puerperal fever mortality in the maternity clinic after the hand-washing regimen was imposed—was quick and dramatic. In the preceding years, the mortality incidence in the clinic had been as high as 18.3 percent. One year after he implemented the hand-washing regimen, it dropped to 1.27 percent—a massive and glaring signal of benefit. Even so, many academic eminences of his day insisted that Semmelweis's deductions were erroneous. His story is a notable example of academic obtuseness parodied by the "Parachute Paradigm." After a couple of observations, it becomes apparent that people who jump out of airplanes with parachutes have better outcomes than those who jump out of airplanes without them. An RCT is not needed to validate this observation.

Within the Ivory Tower, seminal work by the epidemiologist and statistician Sir Austin Bradford Hill elucidated multiple ways in which a causal relationship between associations can be determined. He called these "aspects of causal reasoning." Formal experiments are just one of many. Sir Austin was among the first scientists to publish (in 1950) a modern epidemiological study of the link between cigarette smoking and lung cancer. With the benefit of hindsight, we marvel that Sir Austin's insight wasn't already a matter of common knowledge.

To be sure, before he published his paper, other observers had applied basic deductive reasoning to conclude that smoking tobacco probably caused cancer. In 1795, Samuel Thomas von Sömmerring reported incidents of lip cancer among pipe smokers.[133] In 1930, the German physician and researcher Dr. Fritz Lickint published a meta-analysis of 167 reports linking smoking to lung cancer.[134] The reason why it took so long for Dr. Lickint's findings to change public perception and policy is that the tobacco industry waged a massive propaganda campaign to obscure the harmful effects of tobacco.

A key tobacco industry propagandist was the Austro-American Edward Bernays, often called the father of public relations. The nephew of Sigmund Freud, he applied principles of psychology to his propaganda work—first for the US Committee on Public Information during World War I, and later as an advertising executive. His most famous PR campaign was for the American Tobacco Company to promote cigarette smoking among women, for whom smoking had historically been stigmatized. Bernays's first move was to promote smoking as a substitute for eating to appeal to women's desire to be thin. Newspaper and magazine ads displayed photos of beautiful slender ladies smoking cigarettes, along with messages from medical authorities about the benefit of smoking instead of eating sweets.[135]

Far more ambitious was Bernays's 1929 "Torches of Freedom" PR campaign for which he hired scores of women to march in that year's New York City Easter Parade while conspicuously smoking cigarettes. He hired photographers to capture high-quality images of these subjects, which he published all over the world. The imagery of women proudly smoking at a major public event—defying the traditional taboo—proved to be highly effective. Rates of female smoking immediately and sharply increased.[136]

As Bernays understood, the most effective propaganda creates desirable associations for what is being promoted, and negative associations with those who oppose it. The association of smoking with women's liberation was positive. Those who opposed it, such as the Anti-Cigarette League of America, were negatively characterized as wanting to keep women oppressed and cloistered at home. As Bernays proved with his PR campaigns and proclaimed in his 1929 book, *Propaganda*, the art of propaganda is highly effective at shaping public perceptions. Even when people are vaguely aware of its role in public affairs, this awareness doesn't diminish propaganda's potency. As he famously noted in his book:

> The conscious and intelligent manipulation of the organized habits and opinions of the masses is an important element in democratic society. Those who manipulate this unseen mechanism of society constitute an invisible government which is the true ruling power of our country.[137]

Professor Risch was acutely aware of the role of propaganda in public affairs, and he suspected that a massive campaign was underway to smear the safety and efficacy of hydroxychloroquine. This campaign employed three prongs of attack: 1) The swift and categorical dismissal of case studies from frontline doctors, 2) The insistence that only large RCTs were sufficient to determine a drug's efficacy, and 3) The conducting of RCTs in severely ill hospitalized patients and then publishing the unfavorable results as though they apply to prehospitalized patients in the disease's early phase.

Most of the time, only major pharmaceutical companies have the resources to conduct large RCTs—namely, on drugs they are formulating for profit. Repurposed drugs such as those used by Drs. Raoult and Zelenko yield no profit, so it's no surprise that the observational/case study method of evaluating them—often referred to as "real-world evidence"—is frequently characterized by the pharmaceutical industry as low quality.

In 1993, a group of British and Scandinavian scientists founded the Cochrane Network to review *all* available evidence of the safety and efficacy of drugs, not just randomized controlled trial data favored by pharmaceutical companies for their new products. Its purpose is to provide independent evaluations, free of pharma industry interests. Cochrane Network research has shown that well-conducted case studies and nonrandomized trials are as valid as evidence derived from well-designed and well-conducted RCTs.[138]

Conversely, RCTs have their own weaknesses and biases. In a 2014 paper published in the *Journal of the American Medical Association*, Stanford Professor John Iaonnides et al. showed that 35 percent of the conclusions of the finest RCTs, assessed by peer review and published in the most respected medical journals, could not be replicated on *reanalysis* of their raw data.[139] An obvious bias in many RCTs is that the researchers conducting them are not impartial because they are the beneficiaries of institutional or political patronage that has an interest in a particular outcome.

Congress officially recognized the validity of "real world evidence" in 2016 when it passed the 21st Century Cures Act. This contained a section on the FDA approval process for new indications for existing drugs in which it explicitly stated that "real-world evidence" was acceptable and should be considered by the FDA.[140] The irony wasn't lost on Professor Risch that, at precisely the moment when this 2016 law had the greatest relevance for public health, it was ignored and flouted.

Professor Risch analyzed five outpatient studies of hydroxychloroquine. These included Professor Raoult's two studies and Dr. Zelenko's case study report. From his analysis, he concluded that within the class of patients who were being addressed—namely, high risk from advanced age, obesity, diabetes, or other chronic conditions—the early administration of the drug produced a large benefit. He published his findings in the *American Journal of Epidemiology* on May 27, 2020.

McCullough read Professor Risch's paper, contacted him, and explained that he was assembling a team to research and write a paper about early treatment. Professor Risch agreed to join. At the same time, he continued to study the data on hydroxychloroquine that emerged in June. On June 5, the British RECOVERY trial results were published in a paper titled "No clinical benefit from use of hydroxychloroquine in hospitalised patients with COVID-19."

The RCT—which was largely funded by the Wellcome Trust and Bill and Melinda Gates Foundation—was conducted on 1,542 hospitalized patients in the UK. The trial was unblinded, meaning researchers knew who got the medication and who didn't. This increased the probability that more severely ill patients got the medication. The study found "no significant difference in the primary endpoint of 28-day mortality (25.7% hydroxychloroquine vs. 23.5% usual care)."[141]

As Professor Risch had already noted, hydroxychloroquine was *not* an appropriate treatment for gravely ill hospitalized patients, for whom it was too late for the antiviral therapy to have any benefit. Additionally, azithromycin and zinc were *not* included in the experimental group therapy. Of additional concern was the dose of hydroxychloroquine administered. As the American doctor and medical researcher Meryl Nass noticed in her review of the trial protocol:

> The HCQ dosing regimen used in the Recovery trial was 12 tablets during the first 24 hours (800mg initial dose, 800 mg six hours later, 400 mg 6 hrs later, 400 mg 6 hours later), then 400 mg every 12 hours for 9 more days. This is **2.4 grams** during the first 24 hours, and a cumulative dose of 9.2 grams over 10 days.[142]

The first 24-hour dose of 2,400 milligrams was 6 times the daily dose recommended by Drs. Raoult and Zelenko. Given that hydroxychloroquine skeptics claimed to be concerned about cardiac toxicity, it seemed very strange indeed that the experimental group was given such a huge dose of the medication. This raised the suspicion of deliberate poisoning to worsen their outcome.

In June, McCullough and Dr. Risch commenced work with twenty other medical and research colleagues. As they all understood, time was of the essence. Their work was of urgent importance for millions of people who had no time to wait. Especially in need of early treatment was the American working class, particularly blacks and Hispanics. Within this stratum of society, it was common for multiple generations to live under the same roof. Many owned mom-and-pop restaurants and shops that were shuttered by lockdowns, leaving them to fend for themselves with few resources.

Others were "essential workers" who maintained critical infrastructure, worked in food production, drove the trucks that delivered goods, and stocked the grocery-store shelves. Those who fell ill were told to stay at home and go to the ER if they could no longer breathe. This was a message without hope, for stories circulated about hospitalized patients who were never seen again, not even in death, because their bodies were cremated. And so, for many workingmen and workingwomen who were struck by severe COVID, it seemed there was no one to turn to for help. However, in the late spring, word began to spread about a doctor in East Dallas who would help. All you had to do was go to her clinic.

CHAPTER 19

A Pilgrimage to East Dallas

On June 30, 2020, Adan Gonzalez sat on his patio in San Antonio, wondering what to do. He was very ill with COVID-19, but from a hard recent experience, he'd concluded that going to hospital was not an option. Suddenly he remembered a YouTube clip he'd seen a few weeks earlier about a doctor in Dallas who treated COVID patients. She was addressing a gathering in Dealey Plaza, on the Grassy Knoll overlooking where JFK was assassinated. He still remembered her speech in which she said that COVID was treatable—that no one had to die from it. He'd found it inspiring and thought she seemed like a smart and passionate lady. Trouble was, he couldn't remember her name. He got out his phone and searched for the video on YouTube with the words "Dallas doctor gives speech about COVID," and there she was: Dr. Ivette Lozano. Thank God YouTube hadn't taken the video down. Not yet. Watching it again, he felt a burst of hope.

Adan was in a bad place. The fifty-six-year-old aircraft mechanic and machinist had endured a lot of hardship in his life, but never anything like this. The trouble had started a month earlier, when his oldest son, Adan Jr., whom he called Junior, had fallen ill with COVID-19. He was a commercial truck driver, and one morning at the warehouse, a guy who worked on his loading dock was violently coughing. Junior reported it to his boss, who assured him the coughing man had just received a negative COVID test.

A few days later Junior started feeling ill. His obesity placed him squarely in the high-risk category of people who should receive early treatment, but like so many Americans at this time, he'd never heard of it. About seven days after his first symptoms appeared, he didn't wake up to go to work, and his fiancée noticed that his lips were blue. She called an ambulance, which took him to a hospital in which he was completely isolated from his family.

Adan found it difficult to get information about Junior's condition. On the few occasions he was able to speak with hospital staff, he asked about treating his son with hydroxychloroquine. One doctor said the hospital wasn't authorized to use the drug. Other staff members dodged the question, apparently pretending they hadn't heard him ask it. Then one day a doctor said he was going to administer remdesivir to Junior. Adan had been reading about treatment. He knew that hydroxychloroquine had been around for a long time and taken by millions, while remdesivir was still a new and experimental drug. Again, this struck Adan as strange.

Adan often wished he could be with his son to give him encouragement. Junior had always been a good and affectionate boy, but he'd needed his father to push him. Adan feared he wouldn't muster the strength by himself, alone in his hospital room. On Wednesday, June 24, a nurse agreed to place Junior's phone on speaker so that Adan could offer him a few exhortations. He had no means of knowing his son's condition. All the nurse would say was that he was awake and could hear his father's voice.

"Son, I'm counting on you to fight this thing," he said. "Your woman and beautiful children need you. Please don't give up. I love you, son. We all love you and are praying for you."

That evening, Junior had a Facetime call with his fiancée. As she later reported the conversation, he seemed in good spirits, smiled a few times, and even flirted with her. Toward the end of the call, something strange happened. A doctor entered the room and said, "I'm afraid you're not going to live much longer. Maybe a day or two." A look of despair appeared on Junior's face. His fiancée was shocked by the doctor's dire statement. Obviously, the patient needed hope and encouragement, not a pronouncement he would soon be dead.

When Adan heard about this call, he became terribly upset.

"The look on his face was so horrible when the doctor said he wasn't going to make it," his fiancée said.

"You don't have to explain that," Adan replied, choking back a sob. "I've seen that look many times." Memories of the boy's dispirited expression on various occasions flashed through Adan's mind. God, he wished he were there to give Junior comfort and encouragement.

This terrible report wasn't Adan's only trouble. He also felt a cold coming on. The next day he felt worse, and the day after that he went to the hospital where his son was staying to request a COVID test. The receptionist handed him a flyer and told him brusquely to leave at once. As he was walking to his car, his cell phone rang. It was a doctor calling to say that Junior had just passed away. It was Friday, June 26, 2020.

Adan was thunderstruck and in a poor mental state to process the news, because his condition was worsening. He tried calling the COVID testing center on the flyer and sat on hold for over two hours before someone answered the phone—and gave him the number of another testing center. The second testing center put him on hold for almost another hour. Finally, he got an appointment for a test. A day later he got the result: *Positive*.

As was the case with his son, Adan's illness took a marked turn for the worse about seven days after his symptoms started. He found it increasingly difficult to breathe. Trying to inhale fully caused him to cough violently. He was definitely headed the way of Junior.

So, there he sat in his backyard, thinking about his son, his own precarious state, and the doctor in Dallas. His wife, Roxane, came out to check on him, and he showed her his phone.

"Watch and listen." She watched the video, and at its conclusion he told her, "I gotta go. I gotta go to Dallas to see this doctor."

"Adan Gonzalez, are you crazy? You're barely breathing. There's no way you can drive all the way to Dallas." This wasn't what he wanted to hear. Roxane couldn't drive him either, because he desperately wanted to avoid giving her the disease.

"Well then, I don't know what to do. All I can say is that I'm not going to the hospital." Roxane became upset and went back into the house. That's when he decided to turn to God.

"Dear God, if this is it, this is it. If this is all you have for me, then that's fine. If not, give me a sign. Give me something. Should I drive to Dallas or should I stay here and die?"

About ten minutes later Roxane returned to the patio, walked up behind him, and put her hand on his shoulder.

"Okay Adan," she said. "Let's go to Dallas." He called Dr. Lozano's office, and the receptionist told him to come as soon as possible. The clinic opened the next morning at 7:30. Adan was exhausted and tried to get some sleep before they hit the road. He went to bed early and awakened at 2:00 a.m. With nothing but the clothes on their backs, a couple of thermoses of coffee, and some oranges, they departed. The drive went well until they were about forty-five minutes south of Dallas, at which point Adan's cell phone rang.

"Adan!" Roxane yelled. "You're swerving all over the place. I think you're falling asleep." They exited the Interstate and parked and rested for about fifteen minutes. Then Adan drank the rest of his coffee, ate an orange slice, and ventured back onto the Interstate. As he approached Dallas from the south, it was a titanic effort to concentrate on the road. At times it seemed that other cars and semitrucks were scarcely an inch away. Then came the bewildering spaghetti of highways branching off from one another to go to different sections of the city. He focused on the Waze voice instructions with all his might.

In the car behind him, Roxane gripped the steering wheel and gasped as Adan's car came perilously close to swiping the concrete medians and other cars. It was going to be a miracle if he made it to Dr. Lozano's office. *My God*, she thought. *This is so hard*. Seemingly against all odds, they made it to the small, freestanding building marked by a sign that read Dr. Ivette Lozano. Roxane thought the building resembled a restaurant, and in fact it had originally been a Pizza Inn.

They weren't the only people who wanted to see the doctor. The parking lot was full, and a multitude, spaced about six feet apart, lined up outside the front entrance. Per the receptionist's instructions, Adan called the reception and told her he'd arrived. She came out to greet him. With the truck window down, he answered her questions while she filled out the forms. She then said she'd call him when his room was ready. So began a long wait. The July sun rose over Dallas and heated the parking lot and Adan's truck, but he couldn't run the AC because it aggravated his cough. He could only sit there with the windows down and sweat it out. Noon came and went, and then around 2:00 p.m., he got the call.

"Please come through the back entrance," the receptionist said. He went through a back door, where a young woman escorted him to a room. A few minutes later, Dr. Lozano entered. She was a nice-looking lady with long, lustrous black hair, flashing brown eyes, and a fine-featured face that he could see clearly because she wasn't wearing a mask.

"Take that mask off right now," she said. "It's not going to help anyone, and it's restricting your breathing. Sit on this table while I listen to your lungs." He did as he was told, and she placed a stethoscope on his back.

"Take a deep breath," she commanded. This he tried to do, but to no avail, and started coughing. She didn't flinch from his cough, and to Adan it was clear that she wasn't at all afraid. She was like a soldier who'd been shot at countless times and for whom combat had become the norm. Unlike so many American doctors practicing medicine by computer or phone, Dr. Lozano learned to recognize the unique pulmonary sounds produced by COVID-19—distinct from any form of pneumonia or bronchitis she'd ever heard before. Because so many doctors refused to see COVID-19 patients in person, they never learned this and many other skills that were acquired by hands-on doctors like Lozano.

"Hang on," she said. "I'll be right back." She ran out of the room and then came back with some supplies, including three shots and a white pill.

"This is the last hydroxychloroquine tablet I have in my office. Take it now." Again, he did as he was told. She gave him three injections—a steroid, vitamin B12, and an antibiotic. She then placed a device for measuring his blood oxygen on his finger and saw that it was 88.

"My goodness, to think that you just drove here from San Antonio," she said. "It's a miracle you made it. But God is good, and he brought you here for a reason." She then pricked his finger to measure his blood sugar and saw it was 140.

"Your blood sugar is too high. It needs to come down below 100. The severity of this disease correlates with high blood sugar and with being overweight. You need to lose around seventy-five pounds."

"Seventy-five!" he exclaimed. "I can't lose that much weight."

"How much did you weigh when you graduated from high school?"

"About 130."

"That's about what you need to be. I was also too heavy before this began, and I had to slim down quickly when I started seeing patients. Because I'm fit, I haven't gotten sick, even though I see sick patients every day. I'm going to give you some instructions, and I want you to do exactly as I tell you. First, you go to this pharmacy to pick up these drugs," she said and handed him a script. "Take your first dose immediately. Second, you must stop eating sugar. I'm going to give you a list of things you can eat. Your wife can pick them up at the grocery store. Third, you need to get an exercise bike immediately and start riding it for an hour per day.

"Where am I going to find an exercise bike?" he asked.

"Walmart has them. Hang on," she said. "I'll be right back."

Adan remained seated on the examination table. Through the thin wall, he heard Dr. Lozano talking to a patient.

"Your blood sugar is still too high," she said, "And you've lost no weight in an entire week. If you don't want to live, then I don't have time for you, because I've got a full house of patients who do."

Dr. Lozano had learned through direct observation that the virus seemed to thrive on blood sugar, which amplified its replication and ignited a fire of inflammation that would consume the lungs, torturing patients with relentless coughing and gasping until they had no strength left to fight.

At this point, many capitulated to mechanical ventilation, for which they were paralyzed, sedated, and sent into oblivion.

Dr. Lozano returned to review her instructions with Adan and told him to come back the next day. He left out the back door, walked to Roxane's car, and explained the drill. She followed him to the pharmacy, where she walked over to his truck to get the script. They spoke for a couple of minutes, at which point she noticed that her husband was breathing easier. *Much easier.*

Maybe it was the anti-inflammatory properties of the steroid and hydroxychloroquine working together. Maybe it was the positive psychological effect of doing something rather than sitting around waiting to die. Perhaps it was a combination of both. Whatever the cause, there could be no doubt he was breathing easier.

"I can't believe this," he said. "I already feel better." At first, she smiled, and then she began to weep—first tears of joy, and then rage. Rage that Junior had been deprived of treatment, and all the lies that had been told about hydroxychloroquine. How difficult and dangerous it had been for them to drive to East Dallas, when they could have easily gotten the same drugs and instructions in San Antonio.

Why did the people in power and the media do this? she wondered. *Was it for power and control? Or was it just for money*? She thought of money when she checked out at the pharmacist. All the drugs combined—hydroxychloroquine, azithromycin, zinc, extra-strength aspirin, and a stomach antacid—cost about $125, with no insurance. She wondered what the hospital had billed for Junior's stay and for his remdesivir.

She thanked the pharmacist with real gratitude, because—as Dr. Lozano had explained to Adan—he was one of the few in Dallas who would fill a hydroxychloroquine prescription for COVID-19. Others would fill it for lupus or rheumatoid arthritis, but not for COVID. Again, this made no sense. Why could patients suffering from these diseases take the medication, while patients suffering from COVID-19 were forbidden from taking it, even if they were desperately sick and nothing else was being offered? Was it safer to die of COVID than to try hydroxychloroquine?

They checked into a motel near the clinic, getting two rooms next to each other, and then drove to Walmart to buy an exercise bike, and then to a grocery store to buy the foods that Dr. Lozano said were permissible, mostly proteins and no sugar. The days that followed were difficult. Adan had little strength, but he did his best to get on the exercise bike for an hour a day.

Every day they went to Dr. Lozano's clinic, where she listened to his lungs and checked his blood oxygen and blood sugar. The former rose and the latter began to drop. On about day five, his blood sugar dropped below 100, and he ceased coughing. By then Roxane had given up her room and moved in with him. Dr. Lozano told her the second room was a waste of money because the hydroxychloroquine Adan was taking would inhibit the virus and its transmission. She made a video about her husband's recovery in sign language for her network of people with deaf family members and posted it on YouTube, which removed it the next day.

"YouTube took down my video about your progress," she told Adan, struggling to fight back the tears. "A video in sign language!"

About two days later, Adan had no respiratory distress. Apart from being very tired, he felt fine. They made a final trip to Dr. Lozano's clinic to thank her and then returned to San Antonio. Back home, Adan continued the exercise and diet regimen she'd drilled into him, and a few months later he was in the best shape of his life. He often thought of Junior and how he might have still been alive if he too had received early treatment.

When McCullough heard about Dr. Lozano, he visited her at her clinic and marveled at the irony that it had become a Mecca of COVID-19 treatment for people who lived all over Texas and even in neighboring states. Dallas is one of the greatest medical centers on Earth. Almost 6,000 doctors practice in Dallas County. Most are affiliated with major hospitals, and their offices are situated on institutional campuses with billion-dollar endowments, huge staffs, and state-of-the-art equipment. Few to none of these hospital-affiliated doctors were seeing COVID-19 outpatients.

And so, hundreds of sick people made pilgrimages to the former Pizza Inn on a commercial strip that Dr. Lozano had converted into a clinic. She was especially well known in communities and churches in East Dallas. McCullough had first heard about her at a Bible study at the Munger Place Church on Bryan Street. Many of its members were convinced she had saved many lives.

He arrived at their appointment a few minutes early and sat in the clinic's small waiting room while she finished her day's work. He overheard her firm but compassionate advice to an obese patient about diet and exercise. She concluded by restating the elements of the treatment plan and then booked the patient's next appointment. McCullough knew she was a good doctor. She cared and wanted to follow up.

He spoke with her for a while, and she showed him a row of files of the COVID-19 patients she'd treated—over 300 and growing. This was hands-on doctoring of the old school. He admired her courage and compassion. He then asked her how she'd formulated her treatment protocol.

"I saw President Trump's press briefing on March 19, and then I saw his tweet about Professor Raoult's study. I knew Professor Raoult had no reason to lie about what he was observing, so I decided to give it a try. Almost all my patients recovered without going to hospital. The few I admitted already had catastrophically low blood oxygen saturation when they came here."

Like Dr. Zelenko, Dr. Lozano had grown up under a communist dictatorship—in her case, Cuba. The experience had instilled in her a love of individual liberty and the conviction that the state should never intrude into the doctor-patient relationship. She understood that dispensing medication must be regulated, but she also knew that prescribing FDA-approved medications off-label had long been a legal and common practice. She considered state meddling in this practice an act of tyranny.

CHAPTER 20

Healers of the Imperial Valley

Because Drs. Zelenko and Lozano were family doctors with small, independent practices—as distinct from holding positions of authority within major medical institutions—they were easy for the media to characterize as eccentric oddballs mixing medicine with politics. The *New York Times* set the tone for its feature on Dr. Zelenko with the headline "Touting Virus Cure, 'Simple Country Doctor' Becomes a Right-Wing Star."

Entirely immune from such dismissive characterizations was Dr. George Fareed of El Centro, California, just north of the Mexican border. After graduating from Harvard Medical School in 1970, Dr. Fareed was a virology researcher at the National Institute of Allergy and Infectious Disease (NIAID) while its current director, Anthony Fauci, was a clinical associate. He then held professorships at Harvard and UCLA before moving to the Imperial Valley. For decades he enjoyed working as a family doctor with close ties to the rural, mostly agricultural community.

With his strong facial features and tanned, weathered skin, he looked like a modern-day Moses who'd been wandering in the California desert for forty days. Following in the footsteps of his father—the famous medical missionary Omar John Fareed—he also went on medical missions to Africa and was also (like his father before him) doctor of the U.S. Davis Cup tennis team. A leading expert on HIV/AIDS, he founded an AIDS clinic in Brawley where he used repurposed drugs to treat the disease. For this service he received the Border Hero Award at the 2008 International AIDS Conference in Mexico City. In 2015, the California Medical Association honored him with the Frederick K.M. Plessner Memorial Award for his practice.[143]

The Imperial Valley has a large population of Mexican migrant workers, and when COVID-19 arrived in March, it became the epicenter of the disease outbreak in California. Dr. Fareed read Professor Raoult's studies and reports on Dr. Zelenko's success, and he concluded they provided sufficient evidence to justify trying the therapy. Working with Dr. Brian Tyson—a young and spirited doctor, always ready for action—at the All Valley Urgent Care clinic in El Centro, he began treating the infected with the same cocktail used by Drs. Zelenko and Lozano. They set up field tents outside the clinic to handle the surge of patients in late March, examining as many as 200 per day. Through their large community network, they got the word out that they would treat anyone who came, and they emphasized "the earlier the better."

Like Drs. Zelenko and Lozano, Drs. Fareed and Tyson encountered resistance from their medical and pharmacy boards. Some pharmacists pretended to fill the entire prescription while leaving hydroxychloroquine out of the bag. This caused treatment delays and obliged the doctors to make additional calls, wasting time they could have spent seeing patients. Other doctors in California spoke disparagingly about them to the press.

Working long hours for days on end, fighting fatigue and incomprehensible resistance from the medical establishment, they persevered with their mission. Much of Dr. Tyson's staff caught the infection, but all were treated and quickly recovered. In September they reached the 1,900 mark of COVID-positive patients to whom they administered the Zelenko Protocol, about 1,000 of whom were at high risk of developing severe disease. Not one of them died or even went to hospital. Likewise, not one experienced an adverse cardiac event. Two patients who arrived at Dr. Tyson's clinic were already in a state of acute respiratory distress that required immediate hospitalization.[144]

In September, Dr. Simon Gold of America's Frontline Doctors invited Dr. Tyson to give a speech on the steps of the U.S. Supreme Court. On October 17, he spoke about his work with Dr. Fareed, treating the rural poor of the Imperial Valley with resounding success. Many of his patients were diabetic, and many were old, but he'd lost none of them. He proclaimed it was time for *all* doctors to treat the disease. Preventing hospitalization and death would enable all Americans to return to school and work.

"We don't need to let fear take away our freedom!" he exclaimed.[145]

A video of Dr. Tyson's rousing speech was posted on YouTube and Facebook and quickly started to go viral. However, before his good news could spread far, the video was taken down. By then a rigorous campaign had gotten underway on social media sites to censor presentations of early treatment. Site administrators had decided that humanity should not hear Dr. Tyson's message. Billions of people all over the world could only isolate at home and wait for a vaccine. If they fell gravely ill before the heralded vaccine arrived, well, that was just too bad.

The censorship of Drs. Fareed and Tyson can only be properly described as a crime, for by the autumn of 2020, they were probably the most experienced doctors in the world in the hands-on treatment of COVID-19. They continued their great work in the months that followed, and a year later, when they recounted their experience in their book *Overcoming the COVID Darkness*, they had successfully treated 7,000 patients. No major newspaper reviewed their book.

CHAPTER 21

"For the sake of our parents"

On July 1, 2020—the same day that Adan Gonzalez sought treatment at Dr. Lozano's clinic—the FDA posted a warning about hydroxychloroquine on its website: *The FDA cautions against the use of hydroxychloroquine or chloroquine for COVID-19 outside of the hospital setting or a clinical trial due to the risk of heart rhythm problems.* Its rationale for the advisory was data purportedly gathered from hospitalized patients.[146]

As Professor Harvey Risch immediately noted, this was a major act of fraud. The proper application of hydroxychloroquine was during the disease's early phase in order to inhibit viral replication, thereby preventing hospitalization. Most hospitalized patients were suffering from a later, more severe stage of the illness that was properly treated with different drugs.

The FDA's warning was also highly misleading about the risk of heart arrhythmia in patients taking hydroxychloroquine and azithromycin. The FDA claimed to draw this conclusion from reports submitted to its Adverse Events Reporting System. However, as Professor Risch noted:

> What the FDA did not announce is that these adverse events were generated from tens of millions of patient uses of hydroxychloroquine for long periods of time, often for the chronic treatment of lupus or rheumatoid arthritis. Even if the true rates of arrhythmia are ten-fold higher than those reported, the harms would be minuscule compared to the mortality occurring right now in inadequately treated high-risk COVID-19 patients. This fact is proven by an Oxford University study of more than 320,000 older patients taking both

> hydroxychloroquine and azithromycin, who had arrhythmia excess death rates of less than 9/100,000 users.[147]

In June and July, Professor Risch analyzed data from seven additional hydroxychloroquine studies that were completed in spite of the fraudulent *Lancet* paper and FDA warning. With each passing week, evidence of the therapy's efficacy and safety grew, so Professor Risch submitted a follow-up letter with augmented data to the *American Journal of Epidemiology*.[148] He then wrote an essay for *Newsweek* titled "The Key to Defeating COVID-19 Already Exists: We Need to Start Using It" in which he presented a summary of the studies and concluded with the exhortation: "For the sake of high-risk patients, for the sake of our parents and grandparents, for the sake of the unemployed, for our economy and for our polity, especially those disproportionally affected, we must start treating immediately."[149]

Two weeks after *Newsweek* published Professor Risch's essay, my 74-year-old mother came down with COVID-19. Her symptoms began with sinus congestion and then the most severe headache she'd ever had. A positive PCR test confirmed she had the disease. The doctor who performed the test and her internist at UT Southwestern told her to isolate at home and go to hospital if she had difficulty breathing.

Still unable to return to my home in Vienna, I was living in her house when she got her diagnosis. When I saw her doctor's e-mail, offering no treatment advice, I asked my mother to write back requesting a prescription of hydroxychloroquine and azithromycin. Her doctor wrote back that, per her hospital's guidelines, she could not prescribe these medications. Reading this e-mail, I became furious.

For five months we'd been barraged with stories of how dangerous COVID-19 is, especially for people over seventy. Most of my mother's friends were terrified of going out, socializing, and even seeing their kids and grandkids. Several times I'd heard public health officials and grandstanding politicians sanctimoniously condemn the "selfish individuals who valued their freedom more than the lives of our mothers." Now we were being told that a brief course of hydroxychloroquine and azithromycin posed a greater risk to my 74-year-old mother's health than COVID-19.

The risk of a fatal cardiac arrhythmia from taking hydroxychloroquine and azithromycin was estimated to be 9/100,000, or 0.01%. The risk of hospitalization from COVID-19 in my mother's age group was possibly as high as 10 percent. A child could weigh the relative risks (9 out of 100,000 vs. 10

out of 100) and come to a more sensible conclusion than the administrators of UT Southwestern.

As anyone who has ever watched a pharmaceutical ad knows, every drug carries some risk of dangerous side effects. Every year in the United States, approximately 56,000 ER visits and 2,600 hospitalizations are caused by Tylenol liver toxicity. Many common drugs have high incidences of adverse reactions. Every year, millions of people suffering from cancer take chemotherapies and immunomodulating medications that carry considerable toxicity and adverse event risk.

Equally vexing was the endless insistence that more evidence of efficacy was needed in order to authorize using hydroxychloroquine. When faced with a high probability of hospitalization if no intervention is made, it is perfectly irrational to demand that the proposed intervention is *certain* to work. Few if any commonly prescribed drugs and procedures for many medical conditions are certain to work. In fact, *nothing* in life is certain. When presented with a dangerous situation, we take action to confront it, even with no guarantee of success. If a fire breaks out in the kitchen, we try to extinguish it. If our dinner companion is choking, we try to perform the Heimlich Maneuver. We don't first demand certainty that our actions will put out the flames or free the obstruction. We simply do our best.

"This is the stupidest thing I've ever seen in my entire life!" I raged. "Doctors didn't think twice about prescribing me hydroxychloroquine when I made month-long trips to Africa. Michael's [my younger brother's] secretary has lupus and she's been taking it for fifteen years. Now your dumbass doctor says you can't take it for five days to treat COVID."

"I don't know what to say," my mother replied. "She's a great doctor at a great hospital, and I've been her patient for years."

"I think she and her hospital suck," I said.

"Well, you're not a doctor and you can't write a prescription."

"No, but I can find one who will," I said defiantly.

"This isn't up to you," my mother said. "I think I should listen to my doctor."

"No, you don't. There's a big-shot Yale professor who has done a deep analysis of hydroxychloroquine, and he says the evidence is overwhelming that it works if you take it early."

I ran to my study and printed Professor Risch's *Newsweek* essay. This, I knew, was my best chance to persuade my mother to try the therapy in spite of her doctor's refusal to prescribe it. Her father had gone to Yale and had always spoken highly of the place. I went back downstairs with the essay and found my mom, lying on her chaise longue and reading a book. On a few occasions she'd remarked that three generations of her maternal ancestors had reclined on this piece of furniture reading books and that, because she had no daughters, I would inherit it. She'd told me this as though it were something for me to look forward to. I studied her and tried to assess her condition. She appeared to have a cold with an occasional cough.

"I want you to read this essay by Professor Risch," I said.

"Put it there on the edge of my bed, and I'll read it in a bit."

"Okay. In the meantime, I'm going to find a doctor to prescribe it." I called around to some friends, asking if they'd heard of any maverick doctors willing to prescribe hydroxychloroquine, and I got a lead. However, after reading Professor Risch's article, my mother called a family friend who is a retinal surgeon. We were in luck, because, like many eye doctors, he'd already concluded that the safety warnings about hydroxychloroquine were total BS. He happily prescribed it and said he intended to take it himself if he caught COVID.

Neither CVS nor Walgreens would fill the prescription, but a small, independent pharmacist would. I fetched it (hydroxychloroquine, azithromycin, and zinc) and delivered it to my mom. She took her first dose with dinner that Tuesday. The next morning, she seemed about the same and took her second dose. At dinner that evening she took her third dose, after which I went for a long walk. I returned home around 8:30 and peeked into her bedroom. She was propped up in bed watching a British detective show on Netflix on her iPad. The screen's glow illuminated her face, and I saw she looked better. The big circles under her eyes were gone, and she laughed at a witty remark made by the show's hero.

"How are you feeling, Mom?" I asked.

"Not bad. I think I'm okay."

The next morning, she seemed even better, and a few days later she was fully recovered apart from fatigue. Despite being around her in the same house for days, I never came down with COVID-19.

Some have suggested she had a mild case and would have enjoyed the same rapid recovery even if she hadn't taken the therapy. Only by not taking it would that have been knowable. I am certain that twenty-four hours after she took her first dose, I observed a dramatic improvement of her appearance, energy, and mood. Thus, I believe there is a high probability that it prevented her from becoming sicker and it hastened her recovery. The experience confirmed my suspicion that the maligning and suppression of hydroxychloroquine was irrational at best, and more likely criminal.

CHAPTER 22

Enlightenment and Censorship

On August 6, 2020, the *American Journal of Medicine* published the paper titled "Pathophysiological Basis and Rationale for Early Outpatient Treatment of SARS-CoV-2 (COVID-19) Infection." Out of the approximately 50,000 papers published so far about the viral infection, this was the first to instruct doctors in how to treat it. The authors, Peter McCullough, Harvey Risch, et al., summarized their work as follows:

> This article outlines key pathophysiological principles that relate to the patient with early infection treated at home. Therapeutic approaches based on these principles include 1) reduction of reinoculation, 2) combination antiviral therapy, 3) immunomodulation, 4) antiplatelet/antithrombotic therapy, and 5) administration of oxygen, monitoring, and telemedicine.[150]

Such were the principles McCullough had followed in treating his frail father. Now every doctor in the world with an Internet connection could read them in a peer-reviewed paper, published in a major academic medical journal. It was a significant milestone in the campaign for early treatment of COVID-19, and for a while it was the most frequently cited paper in all of medicine. Not a single mainstream media outlet covered its publication.

One doctor who read the paper was Ben Marble, an ER specialist in Florida who founded MyFreeDoctor.com—an online consulting service for connecting patients with doctors. Dr. Marble obtained a copyright for the "McCullough Protocol" from the US Copyright Office and made it the standard treatment for COVID-19 patients seeking medical care on

MyFreeDoctor.com, thereby enabling hundreds of thousands of sick people to receive early treatment.

Many of McCullough's fellow doctors read the paper and contacted him to share their experiences and success stories. Most of them were independent doctors with their own offices and labs. This was no coincidence, because the senior administrators of most major hospital centers had told their affiliated doctors that there was no early outpatient treatment for COVID-19 and that any doctor caught prescribing hydroxychloroquine would be subject to disciplinary action.

Initially these top-down directives seemed to apply only to writing prescriptions. However, around the time McCullough's paper was published, hospital administrators also forbade doctors from even speaking favorably about early treatment. He first heard of this sort of gag order from his former colleagues at the Henry Ford Healthcare System. Years earlier, he had worked for Henry Ford and befriended staff members, including the cardiologist William O'Neill, MD, and the infectious disease specialist Marcus Zervos, MD. Both were coauthors of McCullough's paper. On June 2, Henry Ford reported that "Treatment with Hydroxychloroquine Cut Death Rate Significantly in COVID-19 Patients." According to this report:

> In a large-scale retrospective analysis of 2,541 patients hospitalized between March 10 and May 2, 2020 across the system's six hospitals, the study found 13% of those treated with hydroxychloroquine alone died compared to 26.4% not treated with hydroxychloroquine. None of the patients had documented serious heart abnormalities.[151]

The study had been conducted early in the pandemic, when many patients went to hospital as much out of fear as from disease severity. Such patients could have received the medication as outpatients, and the best outcomes were noted in those who started taking the medication *before* they fell gravely ill. The study strongly indicated that disease progression was halted in those who received the medication. Two months after this favorable report was published, Henry Ford's administration ordered Drs. O'Neill and Zervos to stop speaking or writing favorably about the treatment or they would be fired.

How disturbing can this get? McCullough wondered. *They should be celebrating the Henry Ford research accomplishment, not being gagged about it.* This had never happened before with any research discovery or publication. On the contrary, academic hospitals had always bragged about their renowned doctors making discoveries. Dr. O'Neill, who was widely considered the father of modern interventional cardiology, had been lauded for his discovery and promulgation of primary angioplasty for acute myocardial infarction. His work had made the William Beaumont and Henry Ford hospitals proud centers of excellence for heart attack care. *Why would these institutions not want the same for COVID-19?*

Right as McCullough was marveling at the bizarreness of this, he got a call from his own chief medical officer.

"I want you to understand that we are going to follow the NIH and FDA regulatory language, which says don't use hydroxychloroquine. So, when you're out there talking about this, you have to say they are your own views, and not the views of Baylor, Scott, and White."

The officer made it clear that those who formulated the hospital's treatment guidelines were *not* doctors like McCullough, but public health agencies such as the FDA and NIH. Though none of the men and women who ran these institutions were treating COVID-19 patients, they were the ones who determined what drugs could be prescribed.

Initially hydroxychloroquine was the subject of their warnings and negative pronouncements, while ivermectin was largely ignored. However, on August 27, 2020, the NIH finally broke its silence about ivermectin and issued an advisory against using it except in clinical trials. The Institutes' rationale for this recommendation was as follows:

> Ivermectin has been shown to inhibit the replication of severe acute respiratory syndrome coronavirus 2 (SARS-CoV-2) in cell cultures. However, pharmacokinetic and pharmacodynamic studies suggest that achieving the plasma concentrations necessary for the antiviral efficacy detected in vitro would require administration of doses up to 100-fold higher than those approved for use in humans.[152]

This negative assertion was drawn from a study published in the *American Journal of Tropical Medicine and Hygiene* on April 16, 2020, the purpose of which was to urge caution in interpreting the positive results of the April 3 Monash University study.[153] The only claim made by the *Tropical Medicine and Hygiene* authors was that it would probably take an extremely high dose of ivermectin to achieve the same plasma concentration in humans as the in vitro concentration cited by the Monash team and that such a high dose could be toxic. The authors did *not* quantify what constituted a toxic dose of ivermectin, nor did they claim the standard ivermectin dose (for parasitical diseases) was necessarily insufficient for treating COVID-19, nor did they compare the adverse effects of ivermectin toxicity to the risk of hospitalization and death from COVID-19.

The authors also referenced a study by Merck Research Laboratories in 2002 to evaluate the effects of "ivermectin administered in higher and/or more frequent doses than currently approved for human use." The study's authors concluded:

> Ivermectin was generally well tolerated, with no indication of associated CNS toxicity for doses up to 10 times the highest FDA-approved dose of 200 microg/kg. All dose regimens had a mydriatic effect similar to placebo. Adverse experiences were similar between ivermectin and placebo and did not increase with dose.[154]

In other words, the NIH advisory about ivermectin's purported toxicity cited a paper showing that the drug is *not* toxic.

Between April 3 and August 26, numerous studies showing ivermectin's safety and efficacy against COVID-19 were published. Instead of acknowledging the value of these studies, the NIH rotten cherry-picked a seminegative report about dosing concerns and cited it as the rationale for advising against using ivermectin to treat the disease. The NIH advisory was superficial and misleading, but it still erected a major impediment for doctors who wished to prescribe ivermectin for their patients.

CHAPTER 23

"We can beat COVID-19 together."

Shortly after McCullough's paper was published in the *American Journal of Medicine*, his daughter came home from law school for a visit and suggested that he do a YouTube presentation of his paper. She helped him to set up his account and videotape a four-slide PowerPoint presentation of the paper's most salient points that were cut and pasted from the peer-reviewed PDF document. About a week after he posted it, it started to go viral. But then YouTube took it down with no explanation apart from the notice that it violated the platform's "community standards." A doctor wearing a suit and tie, giving a presentation of a peer-reviewed paper published in the *American Journal of Medicine*, violated YouTube's standards of what was safe for public consumption.

Shortly after his video was taken down, McCullough was contacted by Senator Ron Johnson's office. The senator wanted to call a hearing about treating COVID-19, and he invited McCullough to talk about the pathophysiological basis and rationale for early treatment. Professor Harvey Risch was invited to present evidence on the safety and efficacy of hydroxychloroquine as the key antiviral. Dr. George Fareed was invited to testify about his extraordinary success in treating thousands of COVID-19 patients. All three doctors accepted Senator Johnson's invitation, and the conference was scheduled for November 19, 2020.

In the fall of 2020, COVID cases began to tick back up, and McCullough feared the country was headed for a second wave in the winter. He was therefore all the more appalled when the NIH released its COVID-19 Treatment Guidelines on October 9.[155] The nation's premier medical research institution recommended NO treatment of the disease unless the patient was hospitalized and required supplemental oxygen. Only then, the NIH recommended administering remdesivir, plus 6 milligrams per day of the corticosteroid dexamethasone.

Remdesivir's safety and efficacy were as doubtful as ever. On October 15, the WHO announced the results of its large Solidarity trial of remdesivir and concluded "there is currently no evidence that it improves survival or the need for ventilation."[156] The WHO therefore recommended *against* remdesivir "for patients admitted to hospital with COVID-19, regardless of how severely ill they are."

Shortly after the WHO released its study results, the European Society of Intensive Care Medicine also recommended against using remdesivir. Additionally, the European Medicines Agency announced that, judging by data shared by Gilead, "remdesivir has potential side-effects on the kidneys."[157] This finding was also reported by doctors such as Arnaud Hot, head of medicine at Edouard Herriot hospital in Lyon, who documented kidney injuries in some of his patients, prompting him to discontinue its use.[158] This revelation raised thorny questions about the European Union's decision to buy 500,000 courses of remdesivir for over a billion euros shortly before the Solidarity trial results were published.

As for the NIH recommendation of dexamethasone: 6 milligrams per day wasn't even close to being a therapeutic dose. In other words, the so-called "Treatment Guidelines" recommended no treatment at all. Upon reading the document, McCullough concluded it would go down as the greatest expression of therapeutic nihilism in history.

Dr. Paul Marik also read the guidelines and was astonished at how perfectly wrong they were. The recommendation of 6 milligrams per day of dexamethasone was *the wrong drug, the wrong dose, and the wrong duration of Rx*. For decades, pulmonary specialists had recognized methylprednisolone as the standard for treating lung inflammation. Dexamethasone was primarily used for treating brain inflammation. Why did the NIH refuse to recognize this? Same with its recommendation of remdesivir—a drug strongly associated with kidney failure that was likely contributing to the deaths of hospitalized patients. How could the NIH be so wrong?

In the fall of 2020, McCullough saw an increasing body of literature on ivermectin's efficacy in treating COVID-19. He also saw favorable reports on colchicine—an anti-inflammatory drug used to treat gout. He also spoke with Dr. Richard Bartlett, a primary care doctor in Odessa, Texas, who reported favorable results with inhaled budesonide. The corticosteroid had long been used to treat asthma. To be sure, there were no published studies on the efficacy of budesonide in treating COVID-19, but the drug had a well-known safety profile, and it was logical that reducing lung inflammation would reduce respiratory distress.

Dr. Bartlett documented a series of patients who reported immediate relief from the therapy, and none of them required hospitalization. As was the case with other doctors who proposed using a repurposed, generic drug to treat COVID-19, Dr. Bartlett was depicted in the mainstream media as a quack recommending "unproven therapies." About eight months after he made his recommendation, an Oxford University research team conducted a randomized trial (STOIC), from which it concluded: *Early administration of inhaled budesonide reduced the likelihood of needing urgent medical care and reduced time to recovery following early COVID-19 infection.*[159]

In early October, McCullough got to work on an update of his paper on early treatment that would incorporate all the new data that had emerged since his original submission three months earlier. It was a monumental work of scholarship that he worked on day and night to produce. In mid-October, as he was deep into writing the paper, his wife, Maha, came home from work with cold-like symptoms and a fever.

They went to an urgent care to get her tested, but the results wouldn't be available for two days. McCullough happened to have some ivermectin, so he gave her 12 milligrams on day one and another 12 milligrams on day two, at which point her fever subsided and she felt better. The next day McCullough himself developed fever and nasal congestion. Fortunately, it was on a Sunday, and he hadn't been around any patients for a few days. He cancelled the coming week's appointments and went to the urgent care center for a rapid antigen test. The result was positive.

"Is there any treatment for it?" McCullough asked the nurse.

"No, just rest at home and go to the hospital if you become short of breath."

He called Dr. Sabine Hazan, a gastroenterologist in California who owned a company that procured funding and organized clinical trials for a variety of therapeutics.

"I've got COVID-19 and I'd like to contribute to research, if possible," he told her. "Can you get me into a clinical trial for a COVID-19 therapy?" Dr. Hazan enrolled him and Maha and in an open-label study of hydroxychloroquine + azithromycin + a combination vitamin and supplement capsule. The study included wearing a heart monitor for arrhythmias and a pulse oximeter. By the end of that day, they had their prescriptions and heart monitors and began their six-day course. While they were at it, McCullough enrolled himself in a randomized double-blind trial (called COLCORONA) of colchicine vs. placebo. Colchicine is an anti-inflammatory drug used to treat gout and pericarditis. Long after the trial was completed, he learned that he got a placebo.

On treatment day 4, nasal swabs taken from McCullough and Maha tested negative. Treatment day 7 happened to be a beautiful and sunny Halloween in Dallas. By then, Maha was feeling fine, and McCullough fairly well, so he decided to go for a run on his usual loop through Glencoe Park. He felt a little winded but surprisingly well when he reached the park, so he decided to make a YouTube video documenting his recovery. Over the years he'd done a lot of public speaking, but always for scientific presentations. This was the first time he addressed the general public.

> I'm Dr. Peter McCullough. I'm the Vice Chairman of Internal Medicine at Baylor University Medical Center in Dallas, Texas. These are my opinions and not Baylor's. Eight days ago, I developed all of the signs and symptoms of SARS-CoV-2, COVID-19 infection. Six days ago, I tested positive. I had a hot fever, I was short of breath, had nasal congestion, and I felt terrible. I was determined not to be another hospital statistic, so I took action into my own hands. At the advice of a doctor, I took over-the-counter supplements and aspirin. In addition, I took two prescription generic medications for the early treatment of COVID-19. I followed the instructions, and I also played it safe by enrolling in COVID-19 research, because there are no proven drugs that are FDA-approved to treat COVID-19 at home. But with this approach, I'm here to tell you on Halloween, October 31, I feel great, I've overcome the infection, and I'm back to jogging. . . . If you're like me, and you're over age fifty and have some medical problems, I advise that you consider early home treatment, and consult the American Association of Physicians and Surgeons. You can find them online, AAPS, and download the home treatment guide for COVID-19 and discuss treatment options with your doctors. Don't stay at home and sicken for two weeks and end up a hospital statistic. It doesn't have to end that way. We can beat COVID-19 together.

CHAPTER 24

A Devil of a Time

McCullough had always been a political moderate and had voted for Obama in both elections, and Hillary Clinton in 2016. For him, the early treatment of COVID-19 was a matter of healing the sick, and he was astounded at how it became a hotly politicized issue. The political operatives who contacted him in the fall were people who professed to be conservative Trump supporters. This was mildly awkward for McCullough, who'd never been a Trump fan. However, by then it had become clear that treating COVID-19 had become perfectly partisan. People who identified themselves as Republicans were generally open to the concept; people who identified themselves as Democrats tended to dismiss it out of hand. As bizarre as this was, it was just a foreshadowing of the battle that would soon be joined over the COVID-19 vaccines in development.

Key political operatives who introduced McCullough to national politics were Karladine Graves and Jenny Beth Martin. The former was a family doctor in Kansas City; the latter was the cofounder of the Tea Party Patriots and a columnist for the *Washington Times*. In their view, the key antiviral therapy, hydroxychloroquine, had been suppressed because President Trump had endorsed it. Precisely because the drug worked, it would—if widely distributed—greatly mitigate the public health disaster, which would be a big feather in President Trump's cap. His opponents were determined to deprive him of this credit in order to wreck his reelection in November. In other words, the suppression of hydroxychloroquine was the most perfidious electoral gambit in history.

McCullough didn't know if this was true or partly true. What he did know what that something had gone terribly wrong in public affairs that had unleashed a host of primitive and irrational emotions. He sensed a mounting anger and hostility in the hearts and minds of people, and a corresponding capacity for aggression and malevolence. To McCullough, this atavistic spirit was the opposite of the humanity and compassion expressed in the Hippocratic Oath, with its vow to "do no harm" and to treat patients "in accordance with my ability and judgment." Was it just a coincidence that, right as a bad moon was rising over the nation, doctors were being prohibited from adhering to their Hippocratic Oath?

Jenny Beth Martin proposed that McCullough and his colleagues set up a Google Group e-mail for sharing observations and discoveries with other doctors and scientists, thereby expanding everyone's knowledge. Initially the "C-19d Group" consisted largely of Americans, but soon it grew to include doctors and scientists all over the world, as well as journalists. Around this same time, McCullough was contacted by the COVID Medical Network in Melbourne, Australia, founded by Dr. Eamonn Mathieson and organized by Ben Bornstein. They too advocated early treatment, not only to prevent hospitalization and death, but also as an alternative to Australia's lockdown policy, which was one of the most draconian on Earth. They invited McCullough and other early treatment advocates to appear on live Australian television for a discussion with Liberal Party MP Craig Kelly.

The network organized a WebEx conference call with MP Kelly to broadcast on prime-time television. Right as they were about to go live, McCullough's laptop screen turned blue, and he could no longer see the other presenters except for Craig Kelly. He heard voices asking, "Dr. McCullough are you there? We can't see you." With the call faltering from this technical problem, the show cut for a commercial break, and he heard a voice suggest they ditch their computers and move to their cell phones. McCullough attempted this and got the same blue screen on his iPhone.

"It must be the WLAN network," he heard someone say. "Let's switch to the cell signal." Again, McCullough tried this to no avail, so they aborted the call and television segment. Afterward, McCullough called the event organizer, Ben Bornstein.

"I've been doing WebEx for months, and I've never seen a blue screen like that," McCullough said. "I wonder if we were hacked." Ben said he would have IT security look into it, and a week later he got back to McCullough.

"Yeah, we were hacked. It seems that someone really doesn't want us talking about early treatment."

McCullough wondered why such massive resources were being devoted to hindering doctors from talking about treating COVID-19 with safe, repurposed, generic drugs. He discussed his concerns with George Fareed and Harvey Risch.

"I'm afraid our Senate testimony will get hacked if we do it by WebEx," he said. "We appear to be up against some pretty formidable opposition." And so, the three men planned to travel to Washington, DC, on their own dimes in order to give their testimony in person before the Senate Committee on Homeland Security and Governmental Affairs.

McCullough wanted to release his updated paper on early treatment as soon as possible, so he planned to publish it in his own institution's medical journal, the *Proceedings of the Baylor University Medical Center.* He had a good relationship with the editor, Dr. Bill Roberts, who agreed to get the paper into peer review as quickly as possible. The reviewers were brisk in coming back with their comments, and McCullough quickly made the corresponding revisions.

It was critical for the paper to be published before November 19 so that he could cite its National Library of Medicine ID number during his Senate testimony. After weeks of intense work under extreme deadline pressure, the paper was finished. The copy editor checked for degree of overlap with the *American Journal of Medicine* paper, since the new paper represented an update in the treatment protocol, and found it met the standard for being a separate work. Dr. Roberts then submitted the paper to the publisher, Taylor & Francis, and McCullough paid the open access publication fee of $3,500 out of his own pocket.

And then—*silence.* Days went by, and McCullough heard nothing from Taylor & Francis. He sent e-mails requesting clarification and got no reply. Finally, two days before his Senate testimony, he heard back from a senior director who had bad news.

"Dr. McCullough, there's a problem with your paper. It's been pulled from publication, and you need to talk to Dr. Roberts, the editor." Immediately McCullough walked down the hall to his office.

"Dr. Roberts, is there a problem with my paper?" he asked. "It's been accepted, we have a contract, and I paid the publication fee. It's a done deal."

"I just heard back from the National Library of Medicine," Dr. Roberts said. "In my fifty years of publishing, I've never seen this before. The library has pulled your paper from publication because they say it overlaps too much with your first paper. I'm sorry, it can't go forward."

McCullough was thunderstruck by this obvious act of censorship. The deputy editor of the *Proceedings* had expressly checked and approved the degree of overlap with the earlier paper. The National Library of Medicine was an indexing service that processed thousands of submissions every day by means of computer automation. No humans were involved in examining each publication, which is why Dr. Roberts had never seen such a rejection in his decades as an editor. It was the same indexing system that allowed the obviously fraudulent Surgisphere paper on hydroxychloroquine to be published a few months earlier. Truly, it was the most astounding thing McCullough had ever heard of.

Shortly before he was scheduled to depart to Washington, he got a text from his boss, Dr. Kevin Wheelan. *Administration doesn't like your Tubes,* it read, referring to McCullough's YouTube videos, recently restored upon Senator Johnson's request. The next day, Baylor Scott & White's chief medical officer requested a conference call to discuss McCullough's forthcoming Senate testimony. Also on the call's invitation list were three corporate attorneys. McCullough accepted the invitation and then contacted his own attorney, who advised him to record the call and to say as little as possible. McCullough then attended the conference call.

"We've seen your YouTube videos, and we understand you're going to testify in the Senate. We want to remind you to declare that your opinions are your own and do not represent Baylor's. We would also like to see a copy of your remarks."

"I'll make it clear my opinions are my own, but I can't give you a copy of my speech. It's privileged information. I've just submitted it to the Senate, and it's under review."

"We want you to know that we're here for you," said one of the lawyers. "We have government relations attorneys, and we can help you."

At the conclusion of the call, McCullough sent the recording to his attorney, who got right back to him.

"Interesting these guys are telling you to disavow Baylor, but at the same time they're saying they want to help you with your remarks." He pointed this out in a bemused tone, but McCullough could tell that he too found it ominous.

When McCullough told his wife, Maha, about his exchange with the Baylor lawyers, she felt very uneasy, just as she'd felt about the hacking incident the week before.

"Going to Washington with all of this strange stuff going on makes me nervous," she said. "Maybe we should ask for Secret Service protection."

"We can't ask for Secret Service," McCullough said. "We're not in the government."

"I wish we at least understood what you're up against," she said. Maha, the daughter of Palestinian Christians with strong family values, had always been a very steady companion, but these recent incidents had shaken her. McCullough thought it might help to talk things over with their Methodist church pastor, Andrew Forrest. He came over shortly before their departure. It was beautiful late fall afternoon, and they sat on the back patio.

"We don't understand what's happening in our world," McCullough explained. "My YouTube videos about early treatment were taken down, and then my planned WebEx conference with an Australian MP about treatment was hacked. Now my hospital administrators are acting like I'm out of line for accepting a US senator's invitation to testify about the disease. It's as though, for the first time in history, our medical system is opposed to caring for the sick. What on Earth is going on?"

Andrew wasn't at all surprised.

"There are times when evil prevails over good in a large way," he said. "We know from the dark periods of history that this has happened before, and now it's happening again. What you describe is Satan working in the hearts and minds of people, sowing fear, confusion, and anger. All you can do is keeping trying to do good until it turns the tide. For your Senate speech, your message must be joyous and happy and clear, uncluttered by negative emotion. That way you will let the light of God shine forth in this darkness. I believe He has chosen you to do this."

Andrew prayed for God's protection and anointed McCullough and Maha with olive oil from Jerusalem, drawing the cross on their foreheads.

CHAPTER 25

Dr. McCullough Goes to Washington

The flight from Dallas to Washington was empty, as was the Kimpton Hotel near Capitol Hill. McCullough and Maha seemed to be the only inhabitants of the vast edifice. At dinnertime they wandered out in search of a restaurant and had a formidably difficult time finding one that was open. They walked the desolate streets, devoid of humanity, past the Smithsonian and other public buildings that were closed, and finally found a place to eat about three miles from their hotel.

The next morning, November 19, he and Maha got up early and had some time to kill before the hearing. The hotel had no coffee service, so they decided to go to a Starbucks. The concierge said the coffee shops were all closed except for one a couple of miles away. They took an Uber to this shop, which offered only takeout service, so they huddled outside in the cold drinking their coffees. They then caught another Uber to the Dirkson Senate Building, where they rendezvoused with Dr. Risch, his wife, and Dr. George Fareed.

They entered the multistory building, and again, no one was there. No staffers buzzing about, no press—just empty halls echoing their footsteps. They were shown to Senator Ron Johnson's chambers, where they had their first satisfying experience of engaging with another human being since their arrival. McCullough marveled that such a man still existed in Washington. The patrician-looking sixty-six-year-old seemed to step out of a Norman Rockwell painting of an earlier, mythically innocent period of American history. He was warm and engaging, and he obviously cared deeply about his country. He also understood perfectly what Drs. McCullough, Risch, and Fareed were trying to accomplish. He told them the story of his daughter being born with a congenital heart defect. During an operation, while trying to create a baffle in her heart, something went wrong and the surgeons had to make a snap decision while she was on the heart-lung machine, with no guarantee of success. They succeeded in repairing her heart, and she grew up to have a happy life as a nurse practitioner, wife, and mother.

"I know that doctors must use their judgment to make decisions. There were no NIH guidelines for those surgeons. They had to make calls for their patient on the table in front of them. Something has gone terribly wrong so that doctors are no longer using their judgment, and the health agencies are playing all these games with evidence, always insisting on more evidence while Americans are dying by the hundreds of thousands. We've got to overcome this and break the news about early treatment to the public."

They left Senator Johnson's office and made their way to the conference room. As they entered the splendid Federal-style chamber, they saw it was empty. Apart from the Senator Johnson, the only other persons present were the Democrat Minority Chairman Gary Peters from Michigan and the Republican Senator from Wyoming, Mike Envi. No one else on the Homeland Security Committee showed up. Perhaps it was no coincidence that all of them had received large campaign contributions from pharmaceutical companies.

Apart from C-SPAN, no one from the press showed up, either. McCullough, Harvey Risch, and George Fareed took their seats at a massive desk, with McCullough sinking into a low chair that made him feel like a little kid peering over the desktop. He felt self-conscious of looking tired and disheveled. He hadn't been sleeping well, which is common for post-COVID-19 patients, and it occurred to him that morning that he badly needed a haircut.

Senator Johnson opened with obviously heartfelt remarks in which he lamented that the response to COVID-19 had been politicized. As it was a novel infectious disease, it was not fully understood, which had made it necessary to make decisions with imperfect knowledge. Nevertheless, it had long been a basic principle of medicine that—when confronted with a potentially fatal disease—it is best to *try* to intervene as early as possible to stop the disease from progressing.

"Why wasn't this logic applied to the coronavirus?"

This question has baffled me since March, and there's probably not a single explanation. We do know that the coronavirus was politicized and was used as an effective weapon in the presidential election. We also know that some of the suggested therapies included off-the-shelf supplements and off-label uses of widely prescribed drugs. The cost of these therapies is well under fifty dollars, vs. a brand-new drug, remdesivir, that costs over $3,000 dollars, and can only be used in hospital, and therefore does not prevent hospitalization in the first place. Could Big Pharma have played a role in discouraging less costly alternatives? I think the answer seems pretty obvious, even though their methods will no doubt remain obscure.

This hearing is not about promoting one particular therapy over others. . . . But I have to say the absence of any serious NIH study or consideration of hydroxychloroquine, either by itself or in combination with other drugs and supplements, is worth discussing. This has been a drug that has been safely and effectively used to prevent malaria and treat lupus and rheumatoid arthritis for decades. Yet doctors who have had the courage to follow the Hippocratic Oath and use their off-label prescriptions rights to treat patients using hydroxychloroquine have been scorned, and state medical boards have threatened to withdrawal their licenses. The same has happened to pharmacists filling the prescriptions for the drug in some states. Will those using ivermectin and other off-the-shelf, off-label drugs suffer the same fate?

Since the onset of this pandemic, I have publicly advocated for allowing doctors to be doctors, to practice medicine, explore different therapies, and share their knowledge within the medical community, and with the public. I believe international, federal, and state health agencies and institutions have let us down. I fear too many have been close-minded bureaucrats, potentially driven by conflicting interests and agendas.

Tragically, media and social media have failed to ask the right questions and censored what they do not understand. My public advocacy has connected me to doctors and scientists who care, and who compassionately treat their patients in spite of the bureaucratic roadblocks they have encountered. . . . Three of those are here today. To me it is obvious that we should explore every possibility for treating this disease at every stage. Why has there been such resistance to low cost, off-the-shelf therapies that might stop the progression and help keep people out of hospitals and intensive care units? I hope today's hearing can answer that question and help to correct this glaring blunder that has cost far too many lives.[160]

The minority chair, Senator Peters, took the opposite view. Underscoring his fundamentally different position, he wore a heavy black mask in contrast to Senator Johnson's unobscured face. Whereas Senator Johnson spoke of his personal inquiry, Senator Peters seemed to be reading from an official script. McCullough marveled at the irony that, years earlier, as a resident of Michigan, he'd voted for the masked man mumbling muffled platitudes.

The senator spoke of how hard the American people had been hit by COVID-19, with record case counts and deaths. Though these grim numbers indicated the federal policy response had failed, he insisted that FDA and CDC recommendations were still the only way forward, as they and they alone followed the science. In his view, opinions that did not conform to these recommendations were false. He had therefore introduced legislation to create a COVID-19 task force that would reduce the spread of such disinformation and misinformation and thereby save American lives:

> We must also be careful of giving Americans a false sense of security by promoting untested and unproven outpatient remedies. We all want answers that will keep our families healthy and safe, but I'm concerned that many of the treatments that will be discussed today have been presented as panaceas for the coronavirus. It would be irresponsible to give Americans false hopes that these types of treatments will be enough to keep them safe in lieu of other measures that are scientifically shown to slow the spread of coronavirus such as masking and social distancing. Our nation's top scientists must be allowed to do their work without meddling. . . . Unfortunately, this administration has continued to exert pressure on our public health agencies to water down guidance and even to promote unproven treatments, further putting Americans at risk. These actions have also diminished the public's confidence in an eventual coronavirus vaccine and treatments.

There it was again—the same talking point about "giving Americans false hopes." How many times had McCullough heard and read this concern as though "false hope" was a grave danger to Americans? Senator Peters, downcast in his black mask, made his point. The objective of official policy was to give *no* hope at all apart from the heralded vaccine.

It was McCullough's turn to speak. He'd been wondering if he could deliver his 700-word speech within the five-minute allotment. The night before in his hotel room he'd rehearsed it with his wife timing it, and he'd repeatedly gone over the five-minute mark. He needed to increase his tempo while still emphasizing the key points. Now the moment of truth had arrived. He turned his script face down on the Senate desk, looked at Senator Peters, then at Senator Johnson, and began to speak. His self-consciousness vanished, and he felt his voice and confidence strengthen.

He began by pointing out that the country was facing an emergency, with hospitals on the verge of reaching full capacity. It was therefore imperative to develop and deploy early outpatient treatments to prevent people from going to hospital. During the first months of the pandemic, the government and media had focused on contagion control measures such as masking and social distancing. Completely ignored had been early home treatment. He then presented a diagram of the disease's three stages—early viral replication, followed by cytokine storm, followed by microthrombosis. All three of these stages could be treated with a combination of drugs that had long been FDA-approved and widely prescribed for other conditions:

> Doctors in the outpatient communities faced with thousands of patients calling and begging for help have innovated. Dr. Zelenko is one in the middle of the calamity in New York who was an early innovator. I summarized these [treatment protocols] and published them in the American Journal of Medicine . . . and it provides a framework for new agents and drugs to be incorporated in an early ambulatory treatment approach. I've reviewed every report from real world data from doctors who have innovated and faced this problem, and I can tell you that they are achieving rates of hospitalization and death less than three percent for high-risk patients—over fifty years old with multiple conditions. Most doctors can achieve results of less than one percent. With no treatment in the United States right now, an individual over fifty with multiple medical conditions faces a seven percent rate of hospitalization and death. For someone in their eighties, that skyrockets to forty percent. As a doctor I have always treated high risk patients with the best tools available. . . . So, I can tell you right now, I'm not asking for permission to do this, but I am asking for your help. I'm asking the government to organize all agencies to assist doctors with their innovation and their compassionate care of COVID-19 patients at home, because we can prevent hospitalization and death. Thank you.

Right as he finished, he looked at the timer and saw it was at five minutes on the nose—not a second more or less—which astonished him. The jolt of adrenaline from delivering the aggressive speech at exactly 5:00 made his heart pound. *This never happens*, he thought as he remuzzled in the N-95 to listen to Harvey Risch, who was up next.

Professor Risch began by emphasizing that "we need to face this disease head-on and not hide, hoping it will go away":

> In May of this year, I observed that results of a study of a drug used to treat COVID, hydroxychloroquine, were being misrepresented by what I thought at the time was sloppy reporting. We've heard from Dr. McCullough how COVID disease progresses in phases, from viral replication to florid pneumonia, to multi organ attack. Viral replication is an outpatient condition, but the pneumonia that fills the lungs with immune system debris and is life threatening, is hospitalizable. We've also heard how each phase of the disease—each pathological aspect—has to have its own specific treatment. . . . Thus, I was frankly astounded that the studies of hospital treatments were represented as applying to outpatients, in violation of what I'd learned about how to treat patients in medical school. We're now coming to address why, over the last six months, our government research institutions have invested billions of dollars in expensive patent medications and vaccine development, but almost nothing in early outpatient treatment—the first line of response to this pandemic. And it's not that we've lacked candidate medications to study. There have been several. But I think that the early conflation of hospital data with outpatient disease served to imply that the treatment of outpatient disease had been studied and found ineffective. . . . So, I want to reiterate that we are considering the evidence for early treatment of high-risk outpatients to prevent hospitalization and death.

Professor Risch then presented the results of multiple studies showing that hydroxychloroquine—by itself or in combination with other agents—significantly reduced rates of hospitalization and death in high-risk patients. The chief criticism of the studies was that they were nonrandomized, meaning the participants were allowed to choose whether they received the medication or were in the control group that did not. However, as he pointed out, those who chose to take the medication tended to be at higher risk of severe COVID-19 than those who didn't, so this bias bolstered the conclusion of the medication's efficacy. Professor Risch also cited major academic studies—including a massive metastudy by the Cochrane Network in England—showing that nonrandomized studies could yield the same quality of evidence as randomized trials. He also pointed out that many of the most efficacious and widely used drugs in history were established as standard of care without randomized trials.

Next was Dr. George Fareed, whose soft voice seemed to express a gentleness and humility incongruous with his strong, masculine features. As Senator Johnson noted in his introduction, Dr. Fareed had been a virology researcher at the NIAID and a professor at Harvard and UCLA. He'd also successfully treated countless AIDS patients in addition to his award-winning work as a family physician in one of the poorest, most medically underserved regions of the country:

> My experience in treating COVID patients—both in the early flu stage as outpatients and also as hospitalized patients in the ICU—made me determined to prevent the COVID flu from progressing to the horrible, lonely cytokine storm suffering that I saw in the ICU. We accomplished this by what I am presenting here today. Our goal is to treat early. COVID patients are difficult to treat when they get very sick. The Imperial Valley where we work became the COVID epicenter of California in June and July. Since early March, both in my Brawley clinic and in Dr. Bryan Tyson's All Valley Urgent Care Clinic in El Centro, where I also work, over 25,000 fearful people were screened. Over 2,400 were COVID-19 positive, and we treated successfully over 1000 high risk and symptomatic ones . . .
>
> We based our protocol on the great work of Dr. Zelenko and Dr. Raoult. They are our heroes. It was the triple hydroxychloroquine cocktail. HCQ, 3500 milligrams over 5 days, azithromycin, and especially zinc, which is often left out in the studies. The cocktail is best given early, as Dr. McCullough has indicated, within the first five to seven days, when the patient is in the flu stage. The timing of the drug is when the virus is in a very active, maximal replication phase, in the upper respiratory tract, and our goal has been and still is to prevent it from entering the lower respiratory tract, and to prevent hospitalization. And we achieved this in over a thousand patients.
>
> We blend in corticosteroids and prolong the treatment if symptoms warrant, but they generally do not. We use it especially in the high-risk individuals as Dr. McCullough indicated. Those over 50, those with co-morbidities. I used this regimen to treat 31 elderly nursing home residents in an outbreak in June, and 29 recovered fully. The drug works mechanistically through multiple actions. . . . As additional agents become available, they can be added to this cocktail to accentuate its efficacy, and we are routinely now combining ivermectin . . . in a quadruple cocktail with excellent results, since IVM is safe and has a different mechanism of action. This is analogous to the use of multiple agents for HIV treatment. Monoclonal antibodies from Regeneron and Lilly will also be suitable when readily available. The results are consistently

> good, often dramatic, with improvement often within 48 hours. . . . We've seen very few hospitalizations and we've seen not a single negative cardiac event. Our experiences are in line with all of the studies that Dr. Risch mentioned concerning early use of the HCQ cocktail.
>
> And let me be clear: This is only about the science—the science of viral replication, the science of the stages of COVID, and the science of why early treatment works. Our experience has led us to try to communicate our approach, and we think it should be on a national level. We wrote a letter to the President, a letter to Congressmen, a letter to the California health department, and an open letter to Dr. Fauci, and a national plan for COVID-19. As we describe in the national plan, this approach would be part of the solution to the pandemic. If high risk individuals get sick, there is solution for them with early treatment with the anti-viral cocktail. If early treatment becomes widely available, people will become much more confident going back to work and sending their kids back to school. Thank you.

The final witness, who presented a rebuttal to the medical testimony, was Dr. Asish Jha, Dean of Brown University School of Public Health. His testimony, which he gave via WebEx, was a series of assertions against hydroxychloroquine, which he claimed was neither effective nor safe. The only specific study purporting efficacy he mentioned was "a small nonrandomized, nonblinded study of hospitalized patients in France—findings that were later discredited, and the scientist who led that work is now facing disciplinary action." Though Dr. Jha did not state the scientist by name, he was referring to Professor Raoult.

Regarding the question of benefit, Dr. Jha asserted: "There is now clear consensus in the medical and scientific community, based on overwhelming evidence, that hydroxychloroquine provides no benefit in treating COVID-19, including in the outpatient setting." Regarding risk, he asserted: "The FDA issued its EUA for hydroxychloroquine in March. In April, there was a 93 percent increase in related calls to the US Poison Control Centers. These things can hurt."

This final remark was a rhetorical sleight of hand, insinuating that many people had been poisoned by the drug, while he provided no specific numbers, or the average number of related calls prior to April. McCullough responded swiftly and forcefully:

> Senator, I want my testimony to be on record that I think Dr. Jha's testimony is reckless and dangerous for the country. And his comment regarding the

> poison control reporting is exactly what Senator Peters is interested in. You're interested in misinformation regarding COVID. Early in the pandemic when hydroxychloroquine was appropriately used, it kept the March, April, and May curve from skyrocketing. When it was used and approximately 500,000 doses were administered, the poison control centers received—I think the number was 77 additional calls. And when the reviewer looked at it, two-thirds of them took an extra dose and they were concerned. So, it boiled down to 17 cases out of over 500,000 administrations, and yet Dr. Jha holds that up to the American public to scare the public away from a safe and effective therapy for COVID-19.

Professor Risch responded to Dr. Jha's remarks by stating that in fact seven high-quality studies had been performed showing hydroxychloroquine's benefit. Dr. Fareed then reiterated that he had not seen any adverse events in any of the patients he'd treated, and that he'd found it extremely gratifying to see many of them feeling better within twenty-four hours.

Much of the remainder of the hearing consisted of a debate between Dr. Jha and the early-treatment proponents over the efficacy of hydroxychloroquine. Ultimately Dr. Jha made the notable concession that he wasn't particularly concerned about its safety, as all drugs have potentially dangerous side effects. In the final analysis, it came down to the quality of efficacy data, and the gold standard was large randomized controlled trials.

"Senator, I would love it if you would push for better quality randomized trials," he said.

"I have, and they wouldn't do them," Senator Johnson replied. "I've had a direct pipeline to Dr. [Stephen] Hahn [FDA chairman], and I cannot get them to do it, which again begs the question: Why not?"

Senator Johnson's personal engagement in the debate became more passionate.

"As a patient myself, I wanted to get hydroxychloroquine, and I couldn't get it. I think I should have the right to try. . . . And I question that because this cocktail costs about twenty bucks, and remdesivir costs $3,000, if maybe there's a little bit of bias or agenda that's outside of treating a patient."

Later in the discussion, Senator Johnson returned to the difference between protocols established by randomized controlled trials and the judgment of individual doctors treating patients.

"Dr. Fareed, can you talk a bit more in layman's terms about how, as a practicing physician, you approach treating a patient?"

Dr. Fareed replied:

> When you are on the front line and at war—because we are at war in this pandemic—my colleague Dr. Bryan Tyson and I took an aggressive approach, and it involved using the art of medicine, using the principles we have learned. I learned those principles fifty years ago. I was lucky to have gone to Harvard. I wonder if Dr. Jha [who also went to Harvard] actually treats patients by the way he talks.

Dr. Fareed said this cutting remark with an exceptionally gentle voice, in a matter-of-fact tone. A close observer of the C-SPAN video might have noticed a woman in the background (McCullough's wife) turn to whoever was seated to her right and raise her eyebrows.

"Regardless," Dr. Fareed continued:

> I found that what is most important in all of my years of practice is to be adaptable to approaches recommended from other doctors, and examples where there has been a benefit, whether it's modifying an HIV cocktail to be more effective . . . or in this COVID situation we have found this Zelenko treatment to be very effective and reliable.

Senator Johnson asked Dr. Jha if he would like to respond to Dr. Fareed's statement—specifically the question of whether he was treating patients:

> I spent the last twenty years taking care of veterans in the VA healthcare system, and recently switched from the Boston VA to the Providence VA. But as a practicing physician I think a lot about the evidence. I think my patients want me to give science-based treatments. . . . If there is no science and evidence, if I'm going to give an off-label use, I'm very careful about that, because the history of medicine is often one of doing more harm than good, and my first oath as a practicing physician was to do no harm. And so, I'm very much driven by that. And I started off by saying that I believe all three of these gentlemen are very smart and committed and caring and I'd appreciate a similar sort of respect back. I do take care of patients, and I try to do my best by them. Obviously, I'm not perfect, but I try to use science as well as compassion to guide me.

"We appreciate that," said Senator Johnson, "but have you treated any COVID patients?"

"I have not, sir."

Dr. Jha had splendid academic credentials to match his splendid manners, but at this moment he lost a lot of credibility. It was perhaps the equivalent of an aeronautical engineer admitting that he'd never flown in a plane, or a marital counselor admitting he'd never been married.

He implied that Professor Risch—a distinguished epidemiologist twenty years his senior—was categorically wrong in his interpretation of the data. Then he implied that Dr. Fareed's observations as a treating physician were an illusion—that the high-risk patients who received the Zelenko Protocol would have recovered in the same dramatic way without the intervention.

This was probably the most notable moment in the hearing. Since graduating from medical school in 1970, Dr. Fareed had logged fifty years as a medical researcher and treating physician. It would be hard to find a doctor in the entire country with more clinical experience. He testified to the US Senate that he'd successfully treated 1,000 high-risk COVID-19 patients. A few minutes later, a doctor 25 years his junior—one who'd never treated a single COVID-19 patient—asserted that "there is now clear consensus in the medical and scientific community" that a key ingredient of Dr. Fareed's treatment protocol doesn't work. In effect, Dr. Jha told Dr. Fareed to reject the evidence of his own eyes and ears.

Senator Johnson turned his attention from the former to the latter.

"Dr. Fareed, you mentioned that Dr. Zelenko and Dr. Raoult are your heroes. Is it true that Dr. Raoult in France is being prosecuted?"

"Evidently that's the case. That's so tragic and inappropriate, because he's a high-caliber and esteemed doctor in tropical medicine and infectious diseases, and his dedication to his patients is impeccable. He guided me and Dr. Tyson in our area, and he is one of the men who will be accoladed when all of this is over."

"He's being prosecuted right now for using a drug that's been around for sixty-five years, prescribed safely, without EKGs," Senator Johnson remarked, and then turned to McCullough, who raised his hand.

"Dr. McCullough."

"I did a program with Eamon Matthieson at the COVID Medical Network in Australia. To show you how off-kilter the world is, in Queensland, Australia, a doctor will be put in jail for prescribing hydroxychloroquine. . . . In Greece and India, it's in their guidelines to give you hydroxychloroquine, but in Queensland they put you in jail for it."

"I was contacted by a doctor who'd written four prescriptions for hydroxychloroquine," Senator Johnson said. "And she was issued a grand jury subpoena by the Homeland Security Investigation Department, if you can believe that. . . . There is something that has gone wrong here, and I for one intend to get to the bottom of it."

CHAPTER 26

The Empire Strikes Back

The hearing got zero broadcast news attention apart from the C-SPAN recording. Nevertheless, Dr. Jha felt compelled or was incentivized to bash it in an opinion piece for the November 24, 2020, edition of the *New York Times* titled "The Snake-Oil Salesman of the Senate." He opened with likening the event to a contagion.

> There was a super-spreader event last week in the United States Senate. It wasn't the coronavirus, however, that was spreading, but misinformation. . . . The Senate Homeland Security and Governmental Affairs Committee held a hearing about early treatment for COVID-19. Yet instead of a robust discussion about promising emerging therapies or what Congress might do to accelerate such treatments, the conversation was all about the malaria drug hydroxychloroquine. . . . Neither Ron Johnson, the Wisconsin Republican senator, nor his chosen witnesses—three doctors who have pushed hydroxychloroquine—displayed more than a passing interest in evidence. Intuition and personal experiences of individual doctors were acclaimed as guiding principles.[161]

Dr. Jha didn't mention that he himself had focused his Senate remarks on hydroxychloroquine and hadn't mentioned any "promising emerging therapies" apart from vaccines. He also didn't state the names or credentials of the hearing's witnesses or a summary of their findings or experiences. He compared them to the snake-oil salesmen from the frontier past with their advocacy of the drug that President Trump had touted in the spring, implying they were equally lacking in medical sophistication.

"I was called reckless because I pointed to facts that could prevent people from getting the treatment," he wrote, but he didn't state these facts. The online version of his essay hyperlinked the word "reckless" to a similar hatchet job report on the hearing in the *Washington Post.* He claimed the witnesses had expressed a distrust of science and had even "suggested that scientists were part of a 'deep state' conspiracy to deny Americans access to lifesaving therapies." This was, he asserted, "a powerful reminder that not even Congress is immune to toxic conspiracy theories . . ."

Dr. Jha's *New York Times* opinion was, itself, evidence that early treatment of COVID-19 was the subject of a well-orchestrated smear campaign. Why else would such a distinguished academic pen such rank propaganda against his colleagues and their work? That he was personally stung by the revelation that he'd never treated a single COVID-19 patient could only partly account for it.

A possible answer to this question may be gleaned from Dr. Jha's remarks at a January 10, 2017, Georgetown University conference titled "Pandemic Preparedness in the Next Administration." Like the participants at the October 2019 Pandemic Simulation Exercise at Johns Hopkins, Dr. Jha predicted that a devastating pandemic "is going to come at some point." Dr. Fauci, the keynote speaker, made a more precise prediction.

"There is no question that there will be a challenge to the coming administration in the arena of infectious diseases," he proclaimed. "The thing we're extraordinarily confident about is that we're going to see this in the next few years."[162]

As psychiatrist and author Peter Breggin, MD, remarked in his extraordinary book *COVID-19 and the Global Predators: We Are the Prey*, Dr Jha did not speak in a somber tone about the coming devastation. On the contrary, he emphasized that he was excited about the ambitious project of helping the US and other governments, and equally excited about the many pandemic preparation events in Georgetown and Cambridge that lay ahead. The conference was, he said, the "beginning of a journey."[163]

Dr. Jha and his colleagues were animated with the same excitement that denizens of the military-industrial complex would feel at the prospect of a coming war in which they would assume leadership positions. At last, they would be able to deploy all of their forces. With the recognition that the coming war was inevitable, they could call upon the government to allocate far more resources for new technologies, weapons systems, bases, and military organizations. In an atmosphere of such heady excitement, the suggestion of defusing the coming war with diplomacy wouldn't be received with much enthusiasm.

The irony of Dr. Jha's excitement is that, when the pandemic he predicted arrived three years later, he didn't attempt to treat patients or scramble to find consultants to intervene against the disease before it wrecked bodies and imprisoned people in hospitals. Instead, he penned propaganda against hydroxychloroquine and against Drs. McCullough, Risch, and Fareed. Why was the *New York Times* editorial board compelled to publish his misleading account of the Senate hearing? Did the editors even watch the C-SPAN recording of it?

It's not plausible that their motive was a concern about hydroxychloroquine's safety. Dr. Jha himself conceded in his testimony that he wasn't particularly concerned about safety, so why the vast and ceaseless quibbling about whether its efficacy for outpatients had been proven? As Senator Johnson had said in the hearing, *this makes no sense.*

Shortly after McCullough returned home to Dallas, he got another call from a Baylor corporate compliance officer.

"We are concerned about your Senate testimony and the fact that you didn't sufficiently disavow Baylor. Next time you give testimony to a government agency, we want you to check in with us first and let us know what you're going to say."

Shortly after this incident, a doctor's group in Houston submitted an attack piece to the *Dallas Morning News* that resembled Dr. Jha's piece in the *New York Times*. They claimed that McCullough was "pushing the dangerous drug hydroxychloroquine" as though he were a drug-dealing quack. In the comments of the paper's online edition, McCullough wrote that the piece's lead author had never published a single study about COVID-19, and therefore had no standing to claim the drug was dangerous.

Someone in Baylor corporate compliance saw this and sent McCullough a certified letter, reprimanding him for violating the professional behavior clause of his contract by making an ad hominem attack in a public forum. The letter concluded with the notice that if he engaged in such activity again, he would be subject to termination. Upon reading this letter, McCullough sensed that his days at Baylor were numbered.

This was in spite of the growing evidence that he was succeeding in his endeavor to prevent hospitalization and death from COVID-19. Baylor administrators were monitoring his endeavor and didn't approve of it. They apparently had no interest in becoming a "center of excellence" for treating ambulatory patients with COVID-19. To be sure, their lack of ambition wasn't peculiar to Baylor. No hospital in America would ever claim expertise in the ambulatory management of the greatest public health problem of our lifetimes. Not one.

CHAPTER 27

Ivermectin Gets a Hearing

Senator Johnson wanted to follow up the November 19 hearing with a hearing about the growing promise of ivermectin in treating COVID-19. He was already acquainted with Dr. Pierre Kory from his May 6 testimony about treating hospitalized patients with corticosteroids. In the interim, Drs. Kory and Marik, and their FLCCC group, had examined the growing body of literature showing ivermectin's efficacy in treating COVID-19. And so, Senator Johnson invited Dr. Kory to testify once again.

Senator Johnson also sought McCullough's counsel on who to invite to the hearing. McCullough was particularly impressed by Drs. Jean-Jacques and Juliana Cepelowicz Rajter, whose ICON study was published on October 12, 2020, in the *CHEST* journal of pulmonary medicine. He therefore recommended that Drs. Rajter be invited to join Dr. Kory at the next hearing, scheduled on December 8.

The December 8, 2020, edition of the *New York Times*, which hit the newsstands *before* the hearing, featured a critical report on one of the witnesses scheduled to give testimony.[164] The author, Katherine J. Wu, was a science and health reporter. At the time she penned her report, she had no way of knowing what the witnesses would say, but she questioned the scientific validity of their testimony even before it was uttered. Wu focused her report on a cautionary statement that witness Dr. Jane Orient had made about the promised vaccines in an earlier *New York Times* interview.

"It seems to me reckless to be pushing people to take risks when you don't know what the risks are," Dr. Orient said. "People's rights should be respected. Where is 'my body, my choice' when it comes to this?"[165]

Dr. Wu disagreed. Based on the assurances of Pfizer and Moderna, which were based on "early data," she was confident their new vaccines were safe. The title of her report—"No, the Pfizer and Moderna vaccine development has not been 'reckless'"—was misleading. Dr. Orient did not say that "vaccine development has been reckless." Rather, she said it was reckless to pressure people to receive experimental vaccines with limited safety data.

Dr. Jane Orient is executive director of the Association of American Physicians and Surgeons (AAPS). She works closely with Dr. Elizabeth Lee Vliet—former director of the AAPS—to spread the word about early treatment. Their "Guide to Home-Based COVID Treatment," sourced from the McCullough Protocol, is an exceptionally illuminating document and is available free of charge on the Internet. For anxious people all over the world who were desperate for information about what to do if they got sick, the guide was a godsend.

As Senator Johnson remarked in his opening statement, Dr. Orient's earlier statements about the vaccines were irrelevant, because the subject of her testimony, and that of the entire hearing, was early treatment. The *New York Times* report was just one of many in which early treatment was criticized in conjunction with praising the forthcoming vaccines. Critics of early treatment seemed to think it would undermine the rationale for rapid mass vaccination with the new Pfizer and Moderna products.[166]

In his opening statement, Senator Peters, the Democratic minority chair, echoed the *New York Times* report by dismissing the forthcoming testimony before it was delivered. He opened with an assessment of the previous hearing, held on November 19:

> Instead of hearing from expert witnesses about scientific developments in Coronavirus treatments, or how we can improve the pandemic response, the Committee was used as a platform to attack science and promote discredited treatments. The minority witness, an expert in his field, was subjected to personal attacks from other witnesses and from the dais. After testifying, he faced an online harassment campaign unlike any he'd experienced after prior Congressional testimony. . . .
>
> Sadly, it appears that today's hearing will follow the same path, playing politics with public health, and will not give us the information we need to tackle this crisis. . . . The panelists have been selected for their political, not their medical, views. . . . The witnesses have made many harmful, inflammatory statements. Those statements include undermining a COVID-19 vaccine, promoting unproven therapeutics, discouraging commonsense measures to

> stop the spread of the virus like social distancing or masks, and even comparing physicians who support these interventions to supporters of the Nazi regime. . . .
>
> In the coming weeks, we expect that the first Coronavirus vaccines will become available. . . . The Commerce Committee will be holding a subcommittee hearing on how we'll ensure every American can get a vaccine later this week, and I look forward to participating in a robust discussion with scientific and logistic experts.

In response to Senator Peters's opening statement, Senator Johnson said:

> Thank you, Senator Peters. I hope that anyone who heard your opening statement and anyone who read the [news] articles prior to this hearing will actually go back and look at the video of the previous hearing and listen to the witnesses . . . and understand how far from reality the ranking member's statement really was.

Dr. Jean-Jacques Rajter's testimony, delivered by WebEx, offered a powerful message of hope for patients hospitalized with COVID-19 and their families:

> In early April, Dr. Kylie Wagstaff and a team of researchers at Monash University in Australia published data indicating that ivermectin was effective at reducing COVID-19 viral loads by 5000 fold within 48 hours in their cell culture models. The very next day, a COVID-19 patient of mine was rapidly deteriorating. She had gone from room air to nasal cannula, to 50% oxygen over a few hours and was continuing to deteriorate. She would likely be intubated shortly with its associated high mortality.
>
> After discussing her care with her family, I was implored to look for any other possible alternatives to avoid further clinical deterioration. The patient's son was literally pleading with me to find some alternative to save his mother's life. We discussed the results of the in-vitro study using ivermectin. Even though the drug has a great safety record, the patient's son was advised that no dosing trials had been completed. After extensive discussion we agreed that ivermectin had other approved indications and he requested that I attempt use of an approved dosing regimen.
>
> Since no other options were available at the time, informed consent was obtained and ivermectin administered. The patient deteriorated as expected for about 12 more hours, but stabilized by 24 hours and improved by 48 hours.

> Subsequent to this, 2 more patients had similar issues and were treated with the ivermectin based protocol. Based on past experience, these patients should have done poorly, yet they all survived.
>
> This laid the foundation for the ICON study (Use of Ivermectin is Associated With Lower Mortality in Hospitalized Patients With Coronavirus Disease 2019) which was peer reviewed and published in *CHEST*, a major US based medical journal. . . . 107 patients received conventional care and 173 patients received conventional care plus ivermectin. The overall mortality was 25% in the conventional care, whereas it dropped to 15% in the ivermectin treated group. This was a statistically significant difference in favor of ivermectin use. In those patients with severe pulmonary disease at onset, the mortality benefit was even more staggering at 80.7% versus 38.8%. We concluded that further studies were needed to confirm those preliminary findings.

Dr. Rajter, in a careful and pensive voice, then cited twenty-one additional, confirmatory studies showing ivermectin's benefit, which were the subject of a recent meta-analysis published on November 26:

> To summarize, based on the facts as presented above, ivermectin is effective in early disease, late disease, post exposure prophylaxis, and pre-exposure prophylaxis. The response to ivermectin has been well documented. Ivermectin is an oral medication requiring no monitoring. It is safe and has a long track record of such safety. It is inexpensive and widely available.
>
> The US has spent billions of dollars on a multitude of treatment options. My team is ready to proceed with the needed randomized control trials to address any such residual doubt related to ivermectin use. Yet we are unable to proceed due to lack of funding and support. A few hundred thousand dollars may definitively proof or disprove the effectiveness of ivermectin for early treatment with a properly designed and implemented randomized control trial.

Next was Dr. Pierre Kory. Like his close colleague Paul Marik, Kory is a huge bear of man. But whereas Dr. Marik has a soft, genteel way of speaking, Kory is inclined to boisterous expressions of passion and enthusiasm in his New York accent. Socially he's the life of the party, but when he speaks about treating patients, it's with an extraordinary intensity and conviction. He began by pointing out the "two critical deficits in our national treatment response that have made this hearing necessary in the first place":

> Besides the early interest and research into hydroxychloroquine, we can find no other significant efforts to research the use of any other already existing, safe, low-cost therapeutic agents. Seemingly the only research and treatment focus that we have observed on a national scale is with novel or high-cost pharmaceutically engineered products such as remdesivir, monoclonal antibodies, tocilizumab, with all such therapies costing thousands of dollars. This is consistent with conclusions drawn by a physician consulting to Congress about COVID-19 when she concluded, "There is a pervasive problem on the Hill with how we prove the value of a low-cost treatment." Another barrier has been the censorship of all of our attempts at disseminating critical scientific information on Facebook and other social media with our pages repeatedly being blocked.

Dr. Kory then presented the growing body of evidence—including the results of thirteen randomized controlled trials—showing ivermectin's efficacy. He concluded with a plea:

> Recognize that the amount of evidence that I have presented far exceeds the level required for a compassionate use authorization as defined by the FDA. That happened for remdesivir, a drug with far, far less supportive evidence and much, much higher cost. Why can't it happen for ivermectin, given this level of evidence? How many more trials have to be done when our manuscript details results from over twenty with over ten of them randomized?
>
> We are in a pandemic, we are at war, stop pretending this is peacetime where we are conducting business as usual. The NIH must rapidly review the data and make a recommendation. That is not asking for much. The doctors and nurses are tired and getting burnt out. We must get it to stop. I don't know how much longer I can do this, especially knowing that it will all be needless death from here on out, given there is a readily available solution. A solution that cannot be dismissed or ignored.
>
> There is a critical need to inform health care providers in this country and the world. The leadership of our governmental health care agencies has a great responsibility here. All we ask is for the NIH, the CDC, and the FDA to conduct a rapid review of the literature reviewed in this presentation and give guidance to the country's health care providers.

The NIH, CDC, and FDA did not heed Dr. Kory's plea. Most major hospital centers maintained their policy of refusing to administer ivermectin to dying patients, even when the patients and their families begged for it. So began yet another dark chapter in medical history.

CHAPTER 28

Begging for the Wonder Drug

As Michael Capuzzo told the story in his long magazine piece "The Drug that Cracked COVID," Judy Smentkiewicz was an eighty-year-old resident of Buffalo, New York.[167] After working thirty-five years as an office manager for Metropolitan Life and raising two children, she had retired to her small house in the suburbs. A week after Senator Johnson's second Senate hearing, she began preparing for Christmas and looked forward to her two children, Michael and Michelle, visiting her for a few days. However, right after Michael and his wife arrived from Florida, she began to feel unwell. On December 22 she tested positive for COVID. Her kids were devastated and cancelled their Christmas celebration as Judy went into quarantine. A week later, she became short of breath and was rushed to the Millard Fillmore Suburban Hospital. On New Year's Eve she was admitted to the ICU.

It was a terrible moment in which Judy and her children realized they might never see one another again. In the days that followed, the doctors and nurses with whom Michael spoke didn't offer much hope. They said there were no medications for treating COVID-19 approved by federal health agencies apart from remdesivir. This was administered to Judy, but it seemed to have no beneficial effect. On New Year's Eve, as her condition deteriorated, her two children and six of their friends gathered on the street below her hospital window and prayed for her.

Shortly after New Year's Day, Michael received from his mother-in-law a video of Dr. Pierre Kory being interviewed by a reporter for Fox 10 News Now, KSAZ-TV, in Phoenix, Arizona. That morning, Dr. Kory had given his Senate testimony on ivermectin. Michael watched it and was moved by

Dr. Kory's passionate intensity and eloquence. Immediately he called the hospital and told Judy's attending physician that he wanted her to receive ivermectin. The doctor refused on the grounds that it wasn't approved for COVID-19, but Michael refused to take no for an answer, and finally a hospital administrator approved one 15-milligram dose. Less than twenty-four hours later, Judy was taken off the ventilator, and the next day she sat upright in a chair for a Zoom call with her son. She still wasn't out of the woods, and when her heart started racing, she was moved to a cardiac unit, and the hospital refused to give her a second dose of ivermectin. Michael insisted, but the hospital refused to budge.

And so, he contacted his friend and attorney Ralph Lorigo and explained the situation. At the time, Lorigo knew nothing about ivermectin, so he too watched the interview with Dr. Kory and then sued the hospital. New York State Supreme Court Judge Henry Nowak heard the case and ordered the hospital to commence treating Judy with four more doses of ivermectin, per her family doctor's prescription.

The hospital refused to obey the judge's order, which resulted in additional legal wrangling, including another hearing. Finally, the hospital's lawyer agreed to allow Judy's family doctor to administer the drug. He was under the impression it was on hand in the hospital's pharmacy, but when he arrived to carry out his charge, he was told that it would have to be couriered from another facility. This caused another delay. Finally, at 11:00 p.m. that night, the second dose was administered, and she started to improve. Ten days later she walked out of the hospital.

As word spread about Judy's happy outcome, Ralph Lorigo was contacted by countless others in the same situation, and soon his law firm had a new area of practice—trying to force hospitals to administer an FDA-approved, Nobel Prize-winning, WHO "Essential Medication" to dying COVID-19 patients to whom nothing else was offered.

Mr. Lorigo was well suited for the task. The energetic, punctilious attorney and Erie County Conservative Party chairman has a formidable presence, with strong Italian good looks and a penchant for wearing beautifully tailored suit and power ties. Though he specialized in real estate law, he represented his clients seeking ivermectin with great care. A devoted family man with three children and multiple grandchildren, he empathized with the families who sought his help.

To be sure, it wasn't an easy job, because the hospitals fought him tooth and nail, bringing multiple attorneys and expert witnesses to hearings. After a few more successes in which he prevailed and the patients recovered after receiving ivermectin, he received more queries than his staff could handle, so he contacted his friend Beth Parlato and asked her if she would be interested in taking some of the cases.

The 55-year-old attorney and mother of three had served as a judge in a New York State criminal court. Over the course of her career, she'd seen much of the good, the bad, and the ugly, but none of it had prepared her for the grueling path ahead. What she was about to witness would challenge all of her assumptions about the American healthcare and legal systems, and ultimately about human nature itself.

Most of her clients were referrals from the FLCCC, founded by Drs. Marik and Kory. The typical call would come into her office from a desperate husband or wife, daughter or son. Their stories were always the same. A much-loved family member had been languishing in hospital and was now headed for the ventilator and probable death. And though the doctors and nurses stated that the prognosis was poor, the hospital refused to administer ivermectin.

To patients and their families, the situation was incomprehensible. Many of Beth's clients posed a variation of the question: "Mom [or Dad] is declining and is probably going to die, so what's the harm in her trying ivermectin?" Beth was at a loss for an answer. The hospital's policy made no sense, neither as a matter of fact nor law. Many families wondered why "right-to-try" laws didn't apply. Hospital attorneys claimed the "right to try" was only for experimental medications that were not yet FDA-approved. Ivermectin was FDA-approved, just not for the treatment of COVID-19.

Patients and their families found this argument perversely legalistic, but many judges—and all judges elected as Democrats—found it persuasive. Beth argued it was a legal, common, and long-standing medical practice to prescribe FDA-approved drugs off-label. Hospital attorneys retorted that the NIH guidelines for the treatment of COVID-19 did not recommend the off-label administration of ivermectin, and because the NIH was the final scientific arbiter of medical matters in the United States, the hospitals were required to follow its guidelines.

The trouble with the one-size-fits-all NIH guidelines for hospitalized COVID-19 patients was that they didn't work. Almost a year into the pandemic, the United States had the highest COVID-19 death rate of the world's top ten wealthiest nations and was in the top twenty nations with the highest death rates in the world. Approximately 80 percent of hospitalized patients who went on mechanical ventilation died. Also significant was the fact that that on January 14, 2021—in response to Senator Johnson's letter requesting that the NIH review Dr. Kory's presentation of evidence—the NIH dropped its recommendation *against* using ivermectin and adopted a neutral stance. Though far from satisfying for Dr. Kory and his colleagues, the NIH neutral stance at least gave doctors greater leeway to exercise their clinical judgment about the drug.

To make matters even more confusing, healthcare professionals were provided with broad legal immunity by the federal PREP Act (Public Readiness and Emergency Preparedness) of 2005. This authorized the secretary of Health and Human Services to deploy a wide array of "Emergency Countermeasures" in the event of an infectious disease outbreak. When invoked by the secretary of Health and Human Services, the PREP Act provides immunity for the "manufacture, testing, development, distribution, administration, and use of covered countermeasures." On February 4, 2020, HHS Secretary Alex Azar declared COVID-19 an emergency and invoked the PREP Act.

The CARES Act of March 27, 2020, also provided immunity for healthcare workers treating COVID-19 patients. Additional immunity was granted by governors' executive orders in all fifty states. The governor of New York State, in which Beth was practicing, provided the following immunity:

> **Conduct Covered**: Civil liability for injury or death alleged to have been sustained directly as a result of an act or omission by person(s) covered.
> **Person(s) Covered**: Physicians, physician assistants; specialist assistants; nurse practitioners; licensed registered professional nurses; licensed practical nurses.
> **Conduct Not Covered**: Gross negligence.

Many observers who were documenting US healthcare policy with respect to remdesivir wondered if all this liability protection could explain why the new, experimental drug was the hospital standard of care despite numerous red flags raised about its safety. The contrast of this policy with the strict policy *against* administering ivermectin was stunning.

Additionally, all the patients that Beth represented, and their families, stated in writing that they would indemnify the hospitals of liability for any adverse effects apparently caused by ivermectin, and that their primary care physicians would come to the hospital to administer it. Despite these multiple provisions of immunity, hospitals were still dead set against giving ivermectin to dying patients.

The hearings were brutal affairs in which hospital attorneys and expert witnesses portrayed Beth's expert witness (on the safety and efficacy of ivermectin) as a delusional quack. Their most common line of attack was that Beth's witness was a lone, eccentric voice in challenging the overwhelming scientific consensus that informed NIH guidelines. This rhetorical strategy ignored the fact that many of mankind's greatest scientific insights were the work of individuals who challenged the orthodoxy of their day. The growing body of evidence, including RCTs, cited by Beth's witness was dismissed by hospital experts with the assertion that the evidence was "low quality." Thus, the judge was presented with opposing expert witness claims about the evidence, only with the hospital's witness also claiming he had "scientific consensus" and therefore the NIH on his side.

Beth tried to argue that the patient retained sufficient bodily autonomy to decide if he or she wished to take an FDA-approved drug off-label. The hospitals' attorneys retorted that hospital patients had never had the right to decide their treatment, and that granting it with ivermectin would set a terrible precedent, opening a Pandora's box of future patients demanding treatments after hearing anecdotes about their efficacy. Beth regarded this argument as another legalistic dodge. Her clients weren't presuming to practice medicine—they were dying men and women, desperately begging for the right to try an FDA-approved drug as a last and only hope when nothing else apart from remdesivir was being offered.[168]

The hospitals claimed total sovereignty over the patient—a godlike power over all decisions affecting his life and death, with the patient afforded no say. For most gravely ill patients, the decision of this godlike power resulted in death. Thus, to sick patients and their families, the Lords of Healthcare were neither competent nor compassionate.

CHAPTER 29

An Orgy of Federal Money

Beth often wondered why the hospitals fought so hard to deny patients and their families this wish. In some cases, even when hospitals lost the battle in court, they still refused to carry out the judge's order, thereby risking being held in contempt of court. Beth would never forget the first time this happened to one of her clients. After a grueling battle with hospital attorneys, she finally prevailed and got the court order. The patient's family doctor then wrote the ivermectin prescription, and the family picked it up at the pharmacy and took it to the hospital. However, upon their arrival, a security guard intercepted them in the lobby and escorted them out.

When Beth heard about this, she called the hospital's attorney—a woman about ten years her junior—who, in the ensuing conversation, revealed that she'd advised the hospital not to carry out the judge's order. Beth notified the court, but to her inexpressible dismay, the judge did not hold the hospital in contempt. This set a terrible precedent, and in the following weeks and months, the same hospital and attorney did the same trick at least six times. Not once did a judge hold the hospital in contempt, which signaled to the institution that it was above the law.

"You have blood on your hands," Beth told the attorney in a subsequent phone call.

The entire state of affairs radically departed from common sense, logic, and compassion. Patients and their family members consequently felt dreadful emotions from the perception that their greatest wish—to be given a fighting chance to live—was being denied for unknowable and illogical motives. This produced the hardest feelings imaginable. Being caught in the maelstrom of this conflict often left Beth dispirited and exhausted.

Still, she soldiered on, though soon she concluded that for patients not yet committed to the vent, it was best to pursue a creative, extra-legal solution—namely, smuggling ivermectin into their hospital rooms. Often family members were squeamish about doing this, but among Beth's clients who found a way to do it, the recovery rate was 100 percent. Truly it was a strange new world in which hospital patients were like prison inmates and their families were engaged in the contraband of life-saving medications.

Why were the hospitals so dead set on this policy? Was it just for the "emergency countermeasures" money? As with all attempts to ascertain motive, that of money was a good starting point. According to the US Department of Health and Human Services, the CARES Act created a 20 percent add-on to be paid for Medicare patients with COVID-19. The Act further created a $100 billion fund that was used to financially assist hospitals, "a portion of which will be used to reimburse healthcare providers, at Medicare rates, for COVID-related treatment of the uninsured."[169]

The CARES Act, which was hastily drafted within the context of an emergency, was reminiscent of the Emergency Economic Stabilization Act of 2008 for bailing out Wall Street, only the amount of money created ($2 trillion) for the latter crisis was a much larger sum, equivalent to 10 percent of US GDP. As in the case of financial institutions back then, vast federal funds were created and disbursed to hospitals to assist them in dealing with the emergency. Uncle Sam paid hospitals a fee for performing a COVID test, then another fee for admitting a COVID-positive patient, and then the full daily Medicare (with a 20 percent add-on) hospitalization rate, regardless of the patient's insurance status.

The MD and Minnesota State Senator Dr. Scott Jensen drew attention to this in an April 8 interview with Fox News's Laura Ingraham:

> Right now, Medicare has determined that if you have a COVID-19 admission in hospital, you'll get paid $13,000 dollars. If that COVID-19 patient goes on a ventilator, you get $39,000 dollars, three times as much.[170]

Dr. Jensen was addressing the concern that hospitals were thereby incentivized to code patients as COVID-19 admissions even if they were suffering from other illnesses or injuries. Of additional concern was the CDC's guidance:

> In cases where a definite diagnosis of COVID-19 cannot be made, but it is suspected or likely (e.g., the circumstances are compelling within a reasonable degree of certainty) it is acceptable to report COVID-19 on a death certificate as "probable" or "presumed."[171]

This created the possibility that the death of any patient—including extremely frail people with multiple comorbidities—who also happened to present flu-like or pneumonia symptoms could be attributed to COVID-19, even without a positive test.

In the April 8, 2020, White House press briefing, Dr. Fauci dismissed this concern as "a conspiracy theory," thereby ignoring innumerable, well-documented cases within the American healthcare system of fraud committed as a result of perverse incentives.[172] CARES Act payments were conceived to give hospitals resources to cope with a mass medical emergency. The danger of perverse incentives lay in the fact that hospitals received a 20 percent add-on Medicare rate for all patients, regardless of their age, insurance status, and citizenship, even if the hospital did nothing for the patient but place him in a room with basic supportive care.

On November 2, 2020, the Centers for Medicare and Medicaid Services (CMS) announced it would pay an *additional* 20 percent add-on payment to hospitals that used new FDA-approved drugs for treating COVID-19. The first therapies that were eligible for this bonus were remdesivir and convalescent plasma (extracted from donated blood). The cost of these drugs was covered by the Medicare add-on payment. In other words, Medicare (a US government-funded institution) paid hospitals a 20 percent gratuity on the patient's entire hospital bill (already "enhanced" 20 percent) for using these new drugs.[173] While convalescent plasma supplies were dependent on blood donation, remdesivir had no supply constraints and quickly became the standard hospital treatment for COVID-19. At a Medicare-covered price of $3,100 per treatment course on a drug that cost about $10 per dose to manufacture, this enabled its company, Gilead Sciences, to book $1.9 billion in revenue for the fourth quarter of 2020, and $5.6 billion in 2021, making it the number one hospital drug for the year.[174] Gilead enjoyed these sales during this period in spite of the fact that on November 20, 2020, the WHO recommended *against* using remdesivir.[175]

Over the course of 2020, it also became increasingly apparent that mechanical ventilation did *not* overcome the pulmonary inflammation and thrombosis causing SARS and therefore didn't help patients. On the contrary, the vain effort to increase blood oxygen saturation with ever higher ventilation pressure often injured the patients' lungs, further contributing to their deaths. An obvious strategy for avoiding mechanical ventilation was to treat pulmonary inflammation with methylprednisolone (the established standard) and microthrombosis with anticoagulants. However, neither of these therapies was recommended by the NIH in its October 9, 2020, treatment guidelines.

Could the CARES Act-enhanced payments explain why hospitals insisted on sticking with the NIH's nihilistic treatment guidelines despite their low success rate? To put it more dreadfully: Could it be that hospitals refused to use the FLCCC's MATH+ (with ivermectin) protocol precisely because it could work and thereby imperil the 20 percent enhanced Medicare payment for COVID-19 patients, plus the 20 percent add-on for administering these patients remdesivir, plus the $39,000 payment for mechanical ventilation? It was a horrifying thing to contemplate, but Beth couldn't help wondering.

CHAPTER 30

"The best disguise is the truth."

In his 2019 book, *Code Blue: Inside America's Medical Industrial Complex,* Dr. Mike Magee, an MD and former physician-spokesman for Pfizer, memorably described the corruption of the US healthcare system:

> Cozy relationships and generous gratuities have demonstrated a remarkable ability to corrupt even those we would instinctively put on the side of the angels, including members of the biomedical research community, deans of medical schools, directors of continuing medical education programs, officers at the NIH and FDA, and even seemingly altruistic patient advocacy organizations like the American Cancer Society.
>
> A theologian looking at all this might conclude that American health care has lost its soul. A behavioral economist would point us toward studies showing that the exercise of moral judgment in a business context draws on a completely different cognitive framework from the one we use in making such decisions in our personal lives.[176]

Dr. Magee is one of many observers who has perceived that the American healthcare industry—in its close relationship with US government agencies and funding—closely resembles what President Eisenhower called the "military-industrial complex." In his 1961 Farewell Address, he warned:

> We must guard against the acquisition of unwarranted influence, whether sought or unsought, by the military-industrial complex. The potential for the

> disastrous rise of misplaced power exists and will persist. We must never let the weight of this combination endanger our liberties or democratic processes.

Eisenhower's warning wasn't new. Presidents Washington and Madison also warned about the danger that could arise if the new American Republic allowed the establishment of an organized interest in waging war. Because entanglements and conflicts with foreign powers would necessarily result in massive government spending on the army and navy, this would likely result in organized military interests *seeking* such entanglements and conflicts, even if they in no way benefited the American citizenry.[177]

The inner workings of such complexes, in which participants are motivated by financial rewards, raise a question that goes to the heart of the human condition. Under certain circumstances, can normal and decent people lose their moral judgment to the point of "losing their souls"? As Dr. Magee pointed out, studies have shown that people working together in a profitable enterprise tend to be less constrained by ethical considerations than they are in their dealings with family and friends. Their highly focused goal orientation is perhaps reminiscent of Paleolithic hunters in single-minded pursuit of valuable prey. It seems that when we are engrossed in this mental state, we tend not to think about the negative consequences of our behavior for others outside of the enterprise.

People may be slow to recognize that their organization or community has been corrupted if they benefit from it. As Upton Sinclair famously put it, "It is difficult to get a man to understand something when his salary depends upon his not understanding it." Herein lies the power of patronage. If your patron—i.e., the wealthy man or company that pays your salary and benefits—starts behaving dishonestly, you will probably be reluctant to see and oppose it. This isn't a matter of willful denial. Because your status, sense of purpose, and remuneration are provided by your patron, you may never even think about questioning his conduct.

Amplifying this is what cognitive psychologists call "normalcy bias." When immoral conduct seeps into an organization and goes unopposed for a long time, it may become endemic and therefore seem normal. Americans witnessed this in the corporate scandals of the 2000s, starting with Enron in 2001. This period of financial malfeasance culminated in the great Financial Crisis of 2008, largely caused by the massive sale of fraudulently valued mortgage-backed securities. After the crisis erupted, many wondered why regulatory agencies hadn't seen it coming and stopped it. At root of the problem was "regulatory capture"—that is, incentives for the people who worked for agencies, and especially bond-rating agencies, to turn a blind eye to the corruption they were supposed to be preventing.

A singularly terrifying corruption of a society occurred in Germany during the 1933–45 period, when the country—previously the most advanced and cultured in the world—lapsed shockingly far from civilized norms. Likewise, many intellectuals who prided themselves on their moral and intellectual discernment failed to recognize the criminal nature of the Soviet Union and its allied regimes in Central and Eastern Europe.

Reflecting on this disturbing reality, the Swiss playwright Max Frisch wrote a black comedy titled *Biedermann and the Arsonists,* published in 1953. The play's protagonist, a businessman named Gottlieb Biedermann, reads in the paper that arsonists are afoot in his town. Their modus operandi is to introduce themselves as door-to-door salesmen in need of overnight accommodations and to talk the house owners into allowing them to stay in the attic, where they then set fire to the house. Mr. Biedermann marvels that anyone could be so gullible, and he is confident that he would never be taken in by such an obvious trick.

The arsonists then arrive at his house, and through a combination of apparent normalcy and charm, they persuade Mr. and Mrs. Biedermann to allow them to stay in their attic. In a key scene, one of the arsonists proclaims, "The best disguise, even better than humor and sentimentality, is the truth, because no one believes it." The naive couple can't see what is about to happen to them precisely because it is so out in the open. They mistakenly assume that such perfidy would be cleverly concealed and not hiding in plain sight. The arsonists then set the house on fire, which spreads to the neighboring houses and burns down the entire town. In the final scene Mr. and Mrs. Biedermann are transported to the gates of hell, where they encounter the arsonists, who introduce themselves as the Devil and his companion Beelzebub.

Mr. and Mrs. Biedermann's trip to the gates of hell is suggestive of observations made by the Swiss psychiatrist Carl Jung, who believed that all human beings have a dark side that renders them capable of committing or participating in grossly immoral and even criminal acts. Those who fail to recognize the "Shadow," as he called the dark side of human nature, often fail to recognize that they are participating in a corrupt enterprise. Preferring not to see evil makes them susceptible to it. As Jung put it:

> The shadow is a moral problem that challenges the whole ego-personality, for no one can become conscious of the shadow without considerable moral

> effort. To become conscious of it involves recognizing the dark aspects of the personality as present and real. This act is the essential condition for any kind of self-knowledge.[178]

A dramatic twist of people failing to see what's right in front of them was presented with delightful effectiveness in the 1995 film *The Usual Suspects*. In this iteration, people don't recognize the archvillain because, though he is constantly in their midst, he seems harmlessly inept. He emphasizes his method, and the reality of humanity's fatal delusion, with the famous line, "The greatest trick the Devil ever pulled was convincing the world he didn't exist."

Within the context of current affairs, a similar aphorism may be said of powerful interest groups—namely, "The greatest trick that powerful interest groups ever pulled was convincing the world that everyone who detects and reports their activities is a conspiracy theorist." Only the naivest consumer of mainstream news reporting would fail to recognize that powerful interest groups in the military, financial, and bio-pharmaceutical industries work in concert to further their interests. Their activities cross the line into conspiracy when they commit fraud or other crimes to advance their interests. The term "conspiracy theory" suggests the feverish imaginings of a crackpot mind. This ignores the fact that the United States government prosecutes the crime of conspiracy all the time. As one prominent defense attorney described this reality:

> Any time the government believes that it can allege that two or more individuals were a part of a common agreement to commit the same crime, they will include a charge of conspiracy into the indictment. There is no requirement that all of the members of the conspiracy even know about each other, or even know each other personally.[179]

A person may be charged with conspiracy to commit a crime even if he doesn't know all of the details of the crime. When COVID-19 arrived, the Bio-Pharmaceutical Complex vigorously and exclusively pursued the vaccine solution instead of the early treatment solution. In order to realize their ambition, multiple actors simultaneously waged a propaganda campaign against hydroxychloroquine, ivermectin, and other repurposed drugs.

It's likely that only a relatively small number of these actors knew they were making fraudulent claims about the generic, repurposed drugs and knew they were taking action to impede access to these drugs based on fraudulent claims. These actors were the conspirators. Countless others unwittingly played roles in the conspiracy because they themselves believed the propaganda.

CHAPTER 31

"I'm in a very sensitive position here."

Around the same time that Beth Parlato began her journey as a patient advocate, McCullough finally got his updated paper on early treatment over the finish line. After his first attempt to publish it in the fall was rebuffed, he decided to publish it in a dedicated edition of *Reviews in Cardiovascular Medicine*. Though he'd been the editor for twenty years, for this special edition he contracted a guest editor, who was given full autonomy. By the journal's operating bylaws, every paper always had two reviewers and a final editorial decision. However, for the first time ever, six reviewers were assigned to evaluate the paper, and McCullough had to address all of their comments. It was a prodigious amount of work, but he finally got it done. Among independent doctors and researchers, the paper was quickly read and frequently cited.[180] The mainstream media largely ignored it; a few reporters denigrated it as having appeared in the journal of which McCullough was the editor.

McCullough wasn't the only academic medical doctor who was having trouble disseminating information about early treatment. In Bath, England, an internationally recognized medical researcher named Dr. Tess Lawrie had a profoundly disturbing experience at the beginning of 2021 that would haunt her for the rest of her life. For the previous ten years, she'd headed a firm called Evidence-Based Medicine Consultancy, Ltd., reviewing data for global public health agencies such as the WHO to assist in developing treatment guidelines. The South African native with delicate features has a sterling reputation for competence and integrity, and no conflicts of interest.

On December 26, 2020, she happened to see a YouTube video of Dr. Kory's Senate testimony (before it was taken down by censors). The first thing she thought was how strange it was to see a doctor begging politicians to use a safe old medicine. She listened to his presentation and realized he probably knew what he was talking about. Acutely aware that time was of the essence because COVID-19 hospitalizations were rising, she performed her own rapid review of the studies Dr. Kory referenced. She concluded they contained sufficient evidence to justify trying ivermectin for prophylaxis and treatment, and that the drug was probably a game changer. She also realized that Dr. Kory presented enough evidence to make it unethical to give COVID-19 patients a placebo in a randomized controlled trial, given that the medicine would likely help them. Though she was in favor of conducting more efficacy studies, doing so was no argument against using the drug immediately because it was already known to be safe.

Initially she thought her task would be easy. All she had to do was share her information with the WHO and with the British authorities, and surely, they would be responsive. On January 3, 2021, she sent e-mails marked URGENT to UK Health Minister Matt Hancock and Members of Parliament but didn't hear back from them, nor did she hear back from her colleague at the WHO. On January 7, she made an open video appeal to Prime Minister Boris Johnson, urging him to consider the evidence that ivermectin was a safe and effective drug and informing him she was on standby to work with his Minister of Health. She never heard back from either minister's office.

She then contacted Dr. Kory, and he advised her to get in touch with Dr. Andrew Hill, a senior visiting research fellow in the Pharmacology Department at Liverpool University. An advisor to the Clinton Foundation and the Gates Foundation, since October of 2020 he'd been advising the WHO on ivermectin. Drs. Kory and Marik appeared with Dr. Hill before an NIH advisory board on January 6 to present their findings on ivermectin. Dr. Lawrie contacted Dr. Hill and offered to share with him her Rapid Review of the evidence. They met, shared their data, and agreed to work together to get ivermectin approved as soon as possible.

Dr. Hill agreed to join Dr. Lawrie's team in conducting a Cochrane Systematic Review of the evidence of ivermectin's efficacy in treating COVID-19. To Dr. Lawrie, it seemed he was her most promising ally in getting the drug quickly approved. He was already advising the WHO about it, and already on record concurring with Drs. Marik and Kory about its safety and efficacy. However, shortly after he agreed to collaborate with her, something incomprehensibly strange happened. On January 18, 2021, he published in preprint a paper titled "Meta-analysis of randomized trials of Ivermectin to treat SARS-CoV-2 infection."

Dr. Lawrie would never forget reading this paper for the first time. Dr. Hill's results were exactly what she'd just discussed with him.

> Ivermectin was associated with reduced inflammatory markers . . . faster viral clearance . . . Ivermectin showed significantly shorter duration of hospitalization compared to control. In six RCTs of moderate to severe infection, there was a 75% reduction of mortality.

So far so good, but then she came to his conclusion, which shocked her to the core.

> Ivermectin should be evaluated in larger, appropriately controlled randomized trials before the results are sufficient for review by regulatory authorities.

Upon reading this one sentence, Dr. Lawrie knew it killed her aspiration for rapid approval of ivermectin.[181] Dr. Hill's call for more trials would terribly delay the process. It would take months to fund, organize, and perform a large randomized controlled trial. His conclusion also cast doubt on the evidence assembled by Dr. Kory, and it gave powerful ammunition to the authorities who'd already refused to acknowledge its value. There was also the odd formulation "before the results are sufficient for review by regulatory authorities," suggesting the existing evidence didn't even merit their consideration. Altogether, the author's conclusion sabotaged his study's favorable results.

What on Earth? Dr. Lawrie thought. *People are dying, we're being told that hospitals are overflowing, we have a safe old medicine that could help, yet he's calling for more trials.* Immediately she sent him an e-mail, asking him to retract his paper.

"Your review will cause immeasurable harm," she wrote. He consented to a Zoom conference with her the next day. The video recording shows an increasingly exasperated Dr. Lawrie. Dr. Hill remains remarkably calm as she turns up the pressure on him, though his slumping body language and furrowed brow tell of a conflicted man with a troubled conscience. He also persistently avoids making eye contact with her.

> **Dr. Hill**: I mean, I, I think I'm in a very sensitive position here.[182]

Dr. Lawrie: Lots of people are in sensitive positions; they're in hospital, in ICUs dying, and they need this medicine.

Dr. Hill: Well. . . .

Dr. Lawrie: This is what I don't get, you know, because you're not a clinician. You're not seeing people dying every day. And this medicine prevents deaths by 80 percent. So 80 percent of those people who are dying today don't need to die because there's ivermectin.

Dr. Hill: There are a lot, as I said, there are a lot of different opinions about this. As I say, some people simply . . .

Dr. Lawrie: We are looking at the data. It doesn't matter what other people say. . . . It's absolutely crystal clear. We can save lives today if we can get the government to buy ivermectin . . .

Dr. Hill: Rest assured I'm not going to let this last a long time . . .

Dr. Lawrie: The fact you're saying you're not going to let this last a long time shows you realize the impact of your work. So how long are you going to continue letting people die unnecessarily? What is the timeline? . . .

Dr. Hill: Well, I think . . . I think that it goes to WHO and the NIH and the FDA and the EMEA. And they've got to decide when they think enough's enough.

Dr. Lawrie: How do they decide? Because there's nobody giving them good evidence synthesis, because yours certainly isn't good.

Dr. Hill: Well, when yours comes out, which will be in the very near future . . . at the same time, there'll be other trials producing results, which will nail it with a bit of luck. And we'll be there.

Dr. Lawrie: It's already nailed.

Dr. Hill: No, that's not the view of the WHO and the FDA.

Dr. Lawrie: How long do you think this stalemate will go on for?

Dr. Hill: From my side, with every trial that comes through, we'll be adding it on, I think end of Feb, we'll be there. Six weeks.

Dr. Lawrie: How many people die every day?"

Dr. Hill: Oh, sure, I mean, you know, fifteen thousand people a day.

Dr. Lawrie: Fifteen thousand people a day times six weeks . . .

Dr. Hill: I've got to . . . get as much support as I can to get this drug approved as quickly as possible.

Dr. Lawrie: Well, you're not going to get it approved the way you've written the conclusion. You've actually shot yourself in the foot, and you've shot us all in the foot—everybody trying to do something good. You have actually completely destroyed it.

Dr. Hill: Okay. Well, that's where we'll, I guess we'll have to agree to differ.

Dr. Lawrie: Yeah, well, I don't know how you sleep at night, honestly.

Dr. Lawrie then focused on the question of who was behind Dr. Hill's call for more trials in his paper's conclusion:

> **Dr. Lawrie**: Whose conclusions are those on the review you've done? Who is not listed as an author who's contributed?
> **Dr. Hill**: Well, I mean, I don't really want to get into it. I mean, Unitaid—
> **Dr. Lawrie**: I think it needs to be clear. I would like to know who are these other voices that are in your paper that are not acknowledged. Does Unitaid have a say? Do they influence what you write?
> **Dr. Hill**: Unitaid has a say in the conclusions of the paper. Yeah.
> **Dr. Lawrie**: Okay, so who is it in Unitaid who is giving an opinion on your evidence?
> **Dr. Hill**: Well, I mean, it's just the people there. I don't want to start getting . . .
> **Dr. Lawrie**: I thought Unitaid was just a charity. Is it not just a charity?
> **Dr. Hill**: Yeah.
> **Dr. Lawrie**: So they have a say in your conclusions?
> **Dr. Hill**: Yeah.

With this acknowledgement, saying the word "yeah" twice in a row, Dr. Hill made and held eye contact with Dr. Lawrie for one of the few times in the long call. Unitaid describes itself as "a global health agency engaged in finding innovative solutions to prevent, diagnose, and treat diseases more quickly, cheaply and effectively, in low- and middle-income countries." Top donors listed on its website are France, the United Kingdom, Norway, and the Bill & Melinda Gates Foundation, which chairs its Finance & Accountability Committee. Though Unitaid commissioned Dr. Hill because of his expertise as a virologist and pharmacologist, its unnamed personnel dictated his paper's conclusion.

It's notable that on January 12, 2021—one week before he published his paper in preprint—Unitaid announced it was funding a new $40 million research center on long-acting medicines for infectious diseases at the University of Liverpool.[183] Dr. Hill was a senior research fellow in antiretroviral drug research at the institution that received this enormous investment. He was also an advisor of the Gates Foundation, Unitaid's biggest private donor and the WHO's second largest donor (after the United States government). Dr. Hill was also advising the WHO on ivermectin.

Dr. Hill was paid to perform a meta-analysis of the existing studies on ivermectin and apparently given free rein to conduct it as he saw fit. The "people there" at Unitaid did not require him to change the results. Instead, they instructed him to conclude that more large randomized controlled trials were needed in order to confirm his results. In this way, all the health agencies and mainstream media were given cover to continue stating there is "no evidence" of the drug's efficacy. At the same time, none of the public health agencies or pharmaceutical companies in the UK or the US allocated the funds for a large RCT, thereby ensuring that Dr. Fauci's favorite evidentiary standard would never be met.

On March 14, Dr. Lawrie e-mailed the British government's Therapeutics Taskforce to inquire about its evaluation of ivermectin. The following day the organization replied:

> The Therapeutics Taskforce are monitoring the data on worldwide trials on ivermectin, including the WHO meta-analysis led by Dr. Andrew Hill. We have monitored a collection of small studies which have now been completed and provided some positive signals on the use of ivermectin as a treatment for COVID-19. This is a promising step; however, larger scale studies are still needed to confirm the effectiveness and safety of this treatment.[184]

And so, Dr. Lawrie learned how one sentence—dictated by an unnamed actor in the conclusion of a non-peer-reviewed, preprint paper—killed the approval and rollout of ivermectin. This paper was authored by one scientist, working for the directors of two institutions (Unitaid and the WHO) and funded by the second wealthiest man in the world. As a result of this action by a few powerful persons, millions of COVID-19 patients in the UK, USA, and other countries were discouraged from taking, and even denied access to, a drug taken by 250 million people every year for other illnesses.

CHAPTER 32

The Pied Piper of Science

On April 10, 2020, the mathematician and fund manager Eric Weinstein posted a solo episode of his podcast "The Portal" in which he offered his reflections on Jeffrey Epstein. After vividly describing a 2004 encounter with the notorious financier in Manhattan, Weinstein presented his hypothesis that Epstein was not a money manager, but "more probably a construction of one or more intelligence agencies, interested alternatively in powerful actors and scientists." At around forty-three minutes into the podcast, Weinstein posed an intriguing question:

"Why was Jeffrey Epstein so focused on science?" To elaborate this question, Weinstein offered his general reflections about funding for scientific research:

> For some reason the United States has been losing its appetite for funding high level scientific research and protecting it with academic freedom. . . . And after Vannevar Bush and the Endless Frontier Doctrine forced us to do our blue-sky research inside of universities rather than research institutions dedicated to the purpose, we developed a weird problem. We would no longer be able to pay our scientists. . . . And so, what I have likened this to, is that the United States had something like a Ferrari convertible, and it left the top down so that it filled up with rain, and then it started scrawling "Steal Me," in Mandarin, Farsi, and Russian on its front bonnet. We are not protecting our scientific assets. In fact, when Jeffrey Epstein came back out of prison, I tweeted that Jeffrey Epstein was funding what the American government refused to fund. . . . I was extremely dismayed that we are

> fundamentally leaving this open. We left a niche for such a person to start exploiting us.[185]

An entire stratum of American society, consisting of intelligent people who spent their youth studying science, isn't getting properly paid. Though Weinstein specifically referred to theoretical physicists in the United States, his observation equally applies to life science professionals all over the world. Within the fields of pharmacology, virology, and molecular biology, there is an army of highly trained people whose university incomes are little compared to that of finance professionals, asset managers, and attorneys. As the case of Dr. Andrew Hill illustrates, leaders of the Bio-Pharmaceutical Complex such as Bill Gates have filled the pay vacuum by providing generous resources for thousands of research scientists. In so doing, he has become their patron.

On October 19, 2019, the *New York Times* published a report headlined "Bill Gates Met with Jeffrey Epstein Many Times, Despite His Past."[186] Most notable was the report's references to Epstein's expressed desire, on multiple occasions, to partner with Gates in financing global health technologies. Like Gates, Epstein seemed to have recognized the great power that would derive from creating and controlling technology on which the health of all humans, across all nations, could come to rely.

CHAPTER 33

"Rest in peace, wheezy."

In 2021 a massive propaganda campaign against ivermectin was waged in the US mainstream media. The campaign's key talking point was calling ivermectin a "horse dewormer" while omitting to mention it's also a WHO Essential Medicine for humans and a wonder drug for treating River Blindness and Elephantiasis. The campaign culminated in September 2021, coinciding with a rise of hospitalizations from the Delta variant. Dr. Fauci and the mainstream media falsely claimed that only the unvaccinated were falling ill and compounding their error by taking ivermectin instead of getting the shot.

On the evening of September 7, Late-Night comedy hosts Jimmy Kimmel, Jimmy Fallon, Stephen Colbert, and Seth Meyers all joked about America's multitude of rubes taking "horse dewormer" instead of getting vaccinated. Mr. Kimmel's remarks were the most cutting:

> We've still got a lot of pan-dimwits out there. People are still taking this ivermectin. You know the poison control centers have seen this spike in calls from people taking this livestock medicine to fight the coronavirus. But they won't take the vaccine, which is crazy. . . . Dr. Fauci has said that if the hospitals get any more overcrowded, they're going to have to make some very tough choices about who gets an ICU bed. That choice doesn't seem so tough to me. Vaccinated person having a heart attack, yes, come right on in, we'll take care of you. Unvaccinated person who gobbled horse goo, rest in peace, wheezy."[187]

In addition to his mendacious characterization of ivermectin as "horse goo," Kimmel also misled his audience by proclaiming that "poison control centers have seen this spike in calls from people taking this livestock medicine to fight the coronavirus." He didn't mention that most of the calls were inquiries about mild side effects and that *none* of the callers were hospitalized.[188]

The Late-Night shows were just one segment of the US legacy media that was flooded with stories about poison control calls for ivermectin. This well-known propaganda technique, called "flooding" or "firehosing," broadcasts the same message repetitively and simultaneously over multiple media channels. The Johns Hopkins pandemic simulation, Event 201, featured a conversation about this strategy. Planning board member Matthew Harrington, CEO of Edelman global communications, proclaimed:

> We're at a moment where the social media platforms have to step forward and recognize and to assert that their [claim] to be a technology platform and not a broadcaster is over. They in fact have to be a participant in broadcasting accurate information and partnering with the scientific and health communities to counterweight if not flood the zone of accurate information.[189]

Mr. Harrington did not specify who—that is, which members of the scientific and health communities—would decide what information was accurate and what was dis- and misinformation. Another participant, Jane Halton, chair of CEPI's Board, also spoke favorably about the "flood strategy" using "trusted sources" of information. She too didn't specify who would decide which sources were to be trusted.[190]

A few days after Jimmy Kimmel bid the unvaccinated consumers of "horse goo" to rest in peace, a 71-year-old man name Sun Ng and his wife Sing traveled from Hong Kong to Chicago to celebrate their granddaughter Kaylie's first birthday on September 17. Shortly after they arrived at their daughter's home, Mr. Ng fell ill with COVID-19 and was ultimately admitted to Edward Hospital in Napierville, Illinois.

As in the case of Judy Smentkiewicz, Mr. Ng's condition deteriorated in hospital to the point of requiring mechanical ventilation. His daughter, Dr. Man Kwan Ng, asked the hospital to administer ivermectin to her father. Edward-Elmhurst Health System's administrators refused, which prompted Dr. Ng (who holds a PhD in engineering) to retain the attorney Ralph Lorigo and sue the hospital.

The investigative journalist, Mary Beth Pfeiffer, documented Dr. Ng's struggle to persuade the hospital to administer ivermectin to her dying father. Ms. Pfeiffer is, along with Michael Capuzzo (with whom she is collaborating) one of the few legacy newspaper reporters in the country who has covered the suppression of ivermectin. At the first hearing of Mr. Ng's case, hospital attorneys argued that the court should not order the administration of ivermectin because 1) there could be side effects, 2) ordering ivermectin would violate the hospital's policies, and 3) forcing it to do so would be "extraordinary" judicial overreach. DuPage County Circuit Court Judge Paul Fullerton disagreed and ordered the hospital to allow the Ng family's physician, Dr. Alan Bain, to administer a five-day course.

The hospital appealed this November 1 order, and a second hearing took place on November 5. A hospital doctor estimated that "someone in his [Mr. Ng's] condition, being on a ventilator like that, has a 10 or 15 percent chance of survival." One of the hospital's nurses suggested to Dr. Ng that they "stop all this aggressive care and let [her father] die naturally."

However, as Dr. Ng told the court, she and her mother (Mr. Ng's wife of forty years) "love him dearly and cannot give up on him, even if the defendants have."

Judge Fullerton found the plaintiff's arguments more persuasive.

"I can't think of a more extraordinary situation than when we are talking about a man's life," he said in his November 5 decision. "I am not forcing this hospital to do anything other than to step aside. I am just asking—or not asking—I am ordering through the court's power to allow Dr. Bain to have the emergency privileges and administer this medicine."

Regarding the hospital's claim that ivermectin could cause side effects, Judge Fullerton noted that the drug was known to be generally well tolerated.

"The risks of its [common] side effects"—dizziness, pruritus, nausea, and diarrhea—"are so minimal that Mr. Ng's current situation outweighs that risk by one-hundredfold," he said. This led to a bizarre exchange in which a hospital doctor testified that "the risk is that there is no benefit."

"On the contrary," the judge said, "The possible benefit this court sees is helping save Mr. Ng's life with this drug."

And so, the Ng family prevailed in court, but this didn't stop the hospital from creating another delay with its assertion that Dr. Bain couldn't enter the hospital because he was unvaccinated. This resulted in yet another court hearing in which Judge Fullerton again ruled in the Ng family's favor. With a negative COVID test, Dr. Bain was finally allowed to enter the hospital to start the five-day therapy on November 8. Eight hours after Mr. Ng received his first dose, he was able to undergo a one-hour breathing trial. With each passing day his condition improved, and on day five, he awoke from his medically induced coma and pulled out his endotracheal tube. On November 16 he left the ICU and was breathing on his own, without supplemental oxygen. Two weeks later he left the hospital and went home to his daughter's house.[191,192]

The great lie that ivermectin was merely a "horse dewormer" was perversely bolstered by the fact that, because many doctors refused to prescribe the drug's human preparation, their patients scrambled to obtain veterinary preparations from farm supply shops. Because dosing for ivermectin is a precise formula of micrograms per kilogram of body weight, it wasn't easy for people to figure out the appropriate dose on their own, with no doctors and pharmacists helping them.

Jimmy Kimmel also cracked a joke about ivermectin's manufacturer, Merck, issuing a recommendation against taking the drug for COVID-19.

"Listen, if a pharmaceutical company says, 'Please don't take the drug we're selling,' you should probably listen to them," he said. He didn't mention that Merck had made little money from ivermectin since it went off patent in 1996, or that Merck had recently filed an application with the FDA for the approval of its new, patented antiviral medication, molnupiravir, for which it projected $5 to $7 billion in sales through the end of 2022 (provided sales weren't undermined by competition from a generic drug such as ivermectin).[193]

Between the release of Dr. Andrew Hill's paper calling for more trials on January 18, 2021, and the end of the year, the number of COVID-19 deaths in the United States surpassed the 2020 number. It's impossible to know how many of these 446,000 people could have been saved by ivermectin. The most favorable assessment of the drug's efficacy placed the value at 80 percent, or 356,800 people. Even if that assessment were reduced by half, it still would have amounted to 178,400 lives saved. Most of them were isolated in grim ICUs and committed to lonely deaths on ventilators. Many of their families begged the hospitals to allow their dying moms and dads, husbands and wives, to be allowed to try ivermectin and were denied this right till the end.

CHAPTER 34

"Where's the focus on sick people?"

On the morning of January 28, 2021, McCullough received an e-mail invitation to an end-of-workday meeting with a pair of Health Texas Provider administrators. He'd suspected that such an invitation was coming ever since he'd returned from the Senate hearing. During the month of December, he was assigned to the Morning Report—a long-standing tradition in teaching hospitals in which a senior doctor reviews notable current cases with residents. Departing from tradition, a junior attending physician showed up every morning to accompany him but said nothing, which gave McCullough the impression he was being watched. One morning they reviewed a COVID-19 patient who'd been admitted the night before.

"You understand the disease is treatable," McCullough said to the residents. "I have published on this, and there are in fact outpatient treatments that can prevent hospitalization." As McCullough said this, he noticed the junior attending physician chaperone perk up and look at him as if he'd just struck the chord his escort had been waiting to hear. *Okay, now I know what this is about*, McCullough thought. Like the Holy Office of the Inquisition, Baylor was gathering evidence of heresy—in this case, teaching unorthodox COVID-19 treatments to residents.

His schedule on January 28, 2021, was full of patient consultations. Ironically, his final patient, with whom he had an excellent relationship, was the Baylor Heart and Vascular Hospital's most generous donor. After this consultation, McCullough went to the meeting with a feeling of dread, like he was going to his execution. One of the administrator's was an officious-looking man he'd never met before. As McCullough sat down at the conference table, he pretended to glance at a text message on his phone and activated its voice recorder.

"Well, this is never easy," said the man. "Health Texas Provider has decided not to renew your contract."

"What is your reason for doing this?" McCullough asked.

"There is no reason. We are simply exercising our right not to renew your contract."

"I believe this is an act of retaliation for my US Senate testimony and for my advocacy of treating patients with COVID-19."

"I have no comment," the man replied. "We will pay you for another ninety days, but you are no longer to see your patients, but you are to provide emergency coverage for your patients, and you will be hearing more about termination procedures."

And so, with no due process, one of the most distinguished academic doctors on Baylor's staff—a doctor held in high esteem by his many patients, and himself a generous donor to the institution and its affiliated university—was terminated. That very evening, Baylor staff called his patients and told them their doctor was no longer working for the hospital. McCullough's cell phone lit up with patients calling, wondering what on Earth was going on.

Baylor University was McCullough's alma mater, and though he'd attended UT Southwestern medical school, he'd done his internal medicine rotation at the Baylor University Medical Center in 1987. He had loved the institution and taken great pride in being a part of it. Now he was being cast out of it with no ceremony. The experience felt akin to being ostracized from his family, and it made him terribly sad.

There's nothing like a crisis to show a man who his true friends are. McCullough broke the news to wife, Maha, when he got home from work, and she bore it with perfect equanimity. *God, I married well*, he thought to himself, as he often would in the coming months. That weekend he called Tom Johnston, the CEO of HeartPlace, the largest cardiology group in the Dallas-Fort Worth area. Tom was sympathetic to his plight.

"I've been following you and I know what's going on. I'd be happy to help you out." He graciously hired McCullough to work as an internist and cardiologist at HeartPlace. Though it's a separate company from the Baylor University Medical Center, its offices are located on the same campus. Thus, in spite of being stripped of his Baylor titles, in the months ahead he would still go to the same work location for a small fraction of his previous salary. This added to the surreality of his new life.

As McCullough's attorney pointed out, his contract specified that a nonrenewal was the equivalent of termination, requiring a board meeting with a three-quarter vote, but this wasn't followed. And so, he was able to negotiate a substantial settlement from Baylor. In other words, the administration preferred to abandon contractual due process, lose its world-renowned vice-chair of Internal Medicine, and pay him a large sum to leave instead of allowing him to advocate treating COVID-19, and to exercise his First Amendment right to speak about it in scientific meetings and the media.

Being punished for acting in accordance with his conscience motivated him to redouble his efforts. He gave even more interviews about early treatment on independent news sites. America Out Loud, an independent talk radio station, invited him to do a weekly show about COVID-19. The *McCullough Report* quickly drew a large following. In February, Texas State Senator Bob Hall invited him to attend a March 10 hearing hosted by the Senate Health and Human Services Committee to discuss "Lessons Learned in Pandemic Response."

In the run-up to his testimony, McCullough became aware of yet another disastrous policy response to COVID-19. In spite of the FDA's emergency authorization of Monoclonal Antibodies—and in spite of growing evidence that these products could save lives—it was extremely difficult to find them. McCullough heard stories of sick people calling around to urgent care clinics and doctors' offices, desperately seeking them. He knew of two instances in which patients who were unable to find them ended up dying in hospital.

While health agencies were devoting massive resources to making the vaccines available at every pharmacy, the same agencies were strangely ineffectual at distributing this promising treatment for people who were already sick, even though the manufacturers claimed to have produced enough to meet the demand. On January 11, 2021, NPR produced a segment on this frustrating "treasure hunt" for Monoclonal Antibodies.[194] Two months later, at the time of McCullough's Texas Senate testimony, the situation hadn't improved.

He was grateful for the invitation to address Texas policymakers, and he admired Senator Hall for his valiant effort to give early treatment a fair hearing. However, he found the encounter with state bureaucrats dispiriting in the extreme. It started with his reception at the State Capitol. The forecourt to the splendid granite edifice was draped in a giant army tent in which all those entering were obliged to take a rapid COVID test. This procedure wasn't even required to go to work at the hospital. Were state legislators and their staffs so exceptional to warrant this extraordinary measure for entering the building?

McCullough and his colleague Dr. Richard Urso from Houston were scheduled to give the final testimony. This obliged them to sit through almost six hours of self-congratulatory talks by various HHS division heads about all of their assessments and distributions of masks and logistical preparations for the rollout of the vaccines. Not one of them said a word about helping sick people. By the time it was McCullough's turn to talk, he was irritated, and he quickly went to the heart of the matter.

"Patients who get sick go out to get a diagnosis. Now I'm a COVID survivor, my wife is a survivor, my father in a nursing home is a COVID survivor," he began, the memory of his father imbuing his voice with emotion:

> You get handed a diagnostic test result. It says, here, you're COVID positive. Go home. Is there any treatment? No. Is there any resource I can call? No. Any referral lines, hotlines? No. Any research hotlines? No. That's that standard of care in the United States. And if we go to any one of our testing centers here in Texas, I bet that's the standard of care. No wonder we have had 45,000 deaths in Texas. The average person in Texas thinks there's no treatment.[195]

At this point, McCullough became more emphatic and raised his voice in exasperation:

> There is so much focus on the vaccine. Where's the focus on people sick right now? *This committee ought to know where all of these monoclonal antibodies are*! . . . They ought to have a list of the centers that actually treat patients with COVID-19. . . . I have always treated my patients with something for the virus, and then something for the inflammation, and then something for the thrombosis . . .

After he described this failure to treat the sick, he offered an account of how it had come about.

> So, what happened? Around May [of 2020] it became known that the virus was going to be amenable to a vaccine. All efforts on treatment were dropped. . . . And then Warp Speed went full tilt on vaccine development and there

> was a silencing of any information on treatment. Nothing. Silence. Scrubbed from Twitter and YouTube. Can't get papers published on treatment. . . . So, the program has been to try to reduce the spread of the virus and wait for a vaccine. . . . We've probably had four hours [of talk] on vaccination here, but nothing about treating the sick.

To illustrate how McCullough's suggestions could be implemented, Senator Hall printed a brochure with information about treatment and distributed it to the committee. Later he introduced a bill mandating that testing centers provide a document about treatment to patients who test positive. After he submitted his bill, the Texas Medical Association sent an e-mail to its members stating that it did not support Dr. Hall's resolution on "unproven treatments" and proposed in its place a statewide vaccine registry.

CHAPTER 35

Tucker Carlson Today

A month after McCullough's Texas Senate testimony, he received a call from Tucker Carlson's producer, who invited him for an interview. McCullough accepted, and the interview was scheduled for the May 7, 2021, episode of *Tucker Carlson Today*. The day before he flew to Fort Meyers, Florida, and was met at the airport by a driver working for the studio who gave him a ride to his hotel. McCullough had a splitting headache and asked him if they could stop to pick up some ibuprofen. As they pulled into a convenience store parking lot, the driver said something that surprised him.

"Please wait for a moment in the car while I look around."

"Okay," McCullough replied, a bit confused. The driver got out, looked around, and then opened the door. After McCullough bought the ibuprofen and got back into the car, he asked, "Why did you want me to wait in the car for a minute?"

"I drive a lot of people to interviews with Mr. Carlson. Many have something to say that powerful people don't want said, but you more than anyone. I just wanted to make sure that no one was following us from the airport."

McCullough had no idea how to evaluate this statement. He'd grown accustomed to being censored through electronic interference, but until that moment he'd never thought about being physically hindered from attending an interview. The next morning the driver picked him up at his hotel to take him to the studio, which proved to be in the middle of nowhere. For what seemed like an eternity they drove northwest through a high grassland. Finally, they reached a two-story, sandalwood house with what appeared to be a country store on the ground floor and an office on the top. Next door was a fifties-era motel painted green and white. It looked like the sort of place you'd see on the cover of a Jimmy Buffett album. Was this really the *Tucker Carlson Today* studio?

The driver directed him to enter the green building first, where the receptionist would direct him to a makeup room. There he bumped into Georgia Congresswoman Marjorie Taylor Greene. McCullough had heard she was a zealous Trump supporter. The petite blonde was a delightfully engaging and humorous lady.

"I don't know if I'd vote for you, but I'd love to go to a party with you," McCullough joked, which made her laugh. A pair of scorchingly beautiful women wearing bandage skirts then escorted him across the parking lot—their high heels sinking into the hot gravel—to the studio. Entering it, he saw the preternaturally boyish-looking Tucker Carlson sitting behind a table, wearing a shirt with an open collar, glancing at his interview notes. McCullough's was just one of many interviews on his agenda, so Mr. Carlson got to work.

"You're here because someone sent me last month a video of testimony you gave before the Texas Senate HHS Committee on COVID, and you asked a question that I don't think I'd heard anyone ask, *ever*, and it really stuck out. Here it is?"[196]

The video faded in just before McCullough posed the question, "There's such a focus on the vaccine, where is the focus on people sick right now?"

"Amazing," Tucker said. "Where is the conversation about the treatment of COVID-19?"

As the conversation progressed, McCullough marveled at Carlson's skill and naturalness in leading it. He easily grasped the principles of early treatment. The burning question that he repeatedly and emphatically posed during the 45-minute interview was *why*. Why were doctors being discouraged from treating their patients, and why was information about treatment being censored?

McCullough pointed out that the problem was worldwide, and he gave the example of Queensland, Australia.

"In April, they put on the books as a law: If a doctor attempts to help a patient with COVID-19 with hydroxychloroquine, that doctor will be put in jail for six months."

"What?" Mr. Carlson asked, visibly shocked. "Why?"

"Something is up." They talked a bit about the strange situation in Australia, and then Mr. Carlson asked a great question.

"How could, with such a short period of time, the health regulators of Australia know to the point they codified it in a regulation that hydroxychloroquine is not an effective therapy against COVID-19? How could that be known?"

"They couldn't have known," McCullough replied. "And in fact, there are pieces in the timeline that show something is very wrong in the world, and it is not just in the US. Things are worse in Canada. There are anguishing doctors and nurses in the northern EU and Scandinavia who are witnessing seniors being euthanized."

"You are completely blowing my mind," said Mr. Carlson. "I didn't expect this interview at all. . . . This is really shocking." After touching on McCullough's credentials as a witness, they came to the climactic moment of the interview.

"I am deeply concerned that something has gone off the rails in the world," McCullough said. "It involves science and the medical literature. It involves the regulatory response and populations kept in fear, isolation, and despair."

"This is upsetting, but it's also fascinating I think," Carlson said. "You've alluded a couple of times to something being up is the phrase you used. Can you put a slightly finer point on that? Do you believe that NGOs, the enormous nonprofits that have a lot of sway, it seems, in the public health arena, are exercising influence over COVID policy? Is it some international regulatory body? Is it the WHO? What is this?"

"That's really going to be the goal of investigative journalists to figure this out," McCullough said. He thought to himself that it was going to take a legion of investigators to untangle and delineate what would ultimately be revealed as a massive crime against humanity.

CHAPTER 36

For the Love of Money

Long before COVID-19 arrived, I was a close observer of the pharmaceutical industry. My great-grandparents believed that pharmaceutical labs were at the forefront of progress, relieving suffering and extending life, and for the most part they were right. My paternal great-grandfather owned a large chain of drugstores and was a benefactor of UT Southwestern Medical School. At the time of my birth, my maternal great-grandmother gave me a generous gift of Pfizer stock. She had been impressed by Pfizer's key role in discovering how to mass-produce penicillin during World War II (in which her son was killed in action). Eighteen years later her gift paid for my university education. And then, in 1998, Pfizer received FDA approval to sell Viagra.

Pfizer initially developed the drug to treat high blood pressure and angina pectoris. However, as Pfizer's researchers discovered in clinical trials, the drug was better at inducing erections than managing angina. And so, the company repurposed the drug for erectile dysfunction and launched a massive, global PR and marketing campaign—including seeking moral approval from Pope John Paul II and contracting the war hero and 1996 presidential candidate Bob Dole to be the brand's poster gentleman—that succeeded in making Viagra a blockbuster.[197] Fortunately for me, I still owned a large chunk of Pfizer stock. The price spiked in late 1998 and reached an all-time high in April of 1999. I sold my entire remaining position, which financed my early years as a freelance author, before my first book was published.

So, I learned firsthand why pharmaceutical companies seek to develop blockbuster drugs with fanatical zeal. Formulating a safe and effective new medicine to address a large, unmet need is very difficult and expensive. Performing clinical trials and obtaining FDA-approval is an arduous process that normally takes several years. Thus, if an opportunity for a new blockbuster presents itself, a big drug company like Pfizer will go to extreme lengths to seize it.

Three years after the release of Viagra, I learned that Pfizer was not the entirely respectable company my great-grandmother had believed it to be. I arrived at this realization through my interest in British spy novels. In 2001 I lived in Vienna, around the corner from the Burgkino (Burg Cinema), which still played the 1949 film noir classic *The Third Man* on its big screen every weekend. I spent many a dreary winter Sunday afternoon watching the film. Based on the novella and screenplay by Graham Greene, *The Third Man* is a crime story about Harry Lime—an American running a medical charity in Vienna, who makes a killing selling penicillin on the bombed-out, impoverished city's black market. To increase his profits, he cuts the drug with other substances, thereby destroying its efficacy and causing the patients (including children) to die horribly from their infections.

In the film's most iconic scene, the good guy (played by Joseph Cotton) meets his old friend Harry Lime (played by Orson Welles) on the Giant Ferris Wheel in the Vienna Prater amusement park and tries to appeal to his conscience. At the wheel's apex, the charismatic Harry opens the door, points down to people walking on the ground below, and says:

> Look down there. Would you really feel any pity if one of those dots stopped moving forever? If I offered you twenty thousand pounds for every dot that stopped, would you really, old man, tell me to keep my money, or would you calculate how many dots you could afford to spare? Free of income tax, old man. Free of income tax. . . . Nobody thinks in terms of human beings. Governments don't, why should we? They talk about the people and the proletariat; I talk about the suckers and the mugs. It's the same thing. They have their five-year plans, and so have I.

I sensed that Graham Greene might have based the story on something he'd witnessed or heard about. Doing some research, I learned that Harry Lime was probably based on the British spy Harold "Kim" Philby, with whom Greene worked in British intelligence during World War II. Greene, it seems, discovered that Philby was a Soviet double agent long before he was exposed as such in 1963. Instead of ratting out his friend, he kept it to himself and left the intelligence service in 1944. Several pieces of evidence suggest that when he wrote *The Third Man* a few years later, he based it on his conflicted friendship with Philby.

John le Carré was also fascinated by Graham Greene and Kim Philby, and his thriller *Tinker, Tailor, Soldier, Spy*—one of my all-time favorites—was inspired by the Philby story. His novel *The Constant Gardener* was published in 2001, and I read it with great interest. The story wasn't set in Cold War Europe, but in Kenya, where a British diplomat's wife is brutally raped and murdered. Upon closer examination, the diplomat realizes that she was about to reveal a horrifying crime committed by a pharmaceutical company, which murdered her in order to prevent the exposure.

The novel's plot was reminiscent of a controversial drug trial performed by Pfizer in Kano, Nigeria, in 1996 during a meningococcal outbreak. For the trial of its new antibiotic, trovafloxacin, Pfizer gave 100 children this new drug. The control group of 100 other children received the standard antimeningitis treatment at the time—a drug called ceftriaxone. However, for the control group, Pfizer administered a substantially lower dose of ceftriaxone than the drug's FDA-approved standard.

When the reduced dosing in the control group was discovered, it raised the suspicion that Pfizer did this in order to skew the trial in favor of its new drug. Five of the children who received trovafloxacin died, while six who received the reduced dose of ceftriaxone died.[198] Other children apparently suffered grave injuries from the administration of the experimental antibiotic without their informed consent. The investigation and litigation that ensued was the stuff of a thriller, involving private investigators, bribery, blackmail attempts, and disappearing records. Thirteen years later, in 2009, Pfizer settled out of court with the plaintiffs.

In his author's note, le Carré claimed that nobody and no corporation in the novel was based on an actual person or corporation in the real world.

> But I can tell you this. As my journey through the pharmaceutical jungle progressed, I came to realize that, by comparison with the reality, my story was as tame as a holiday postcard.[199]

In 2009, the same year that Pfizer settled with the trovafloxacin plaintiffs, the *New York Times* reported that a US federal judge assessed Pfizer with the "largest health care fraud settlement and the largest criminal fine of any kind ever" for its illegal marketing of Bextra and three other drugs.[200] The

US Department of Justice was unequivocal in characterizing Pfizer's officers as guilty of grave criminal conduct at the expense of the American public.[201]

Pfizer wasn't alone. Between the years 2002 and 2012, the US Department of Justice filed charges and reached settlements (ranging between $345 million and $3 billion) in 22 major cases of pharmaceutical company fraud, mostly under the US False Claims Act. Violations included off-label promotion of drugs for uses not recognized by Medicare, failure to disclose safety data, paying kickbacks to treating physicians, and making false and misleading statements about safety.

Another notable case was Merck's aggressive marketing of its painkiller Vioxx while knowing and concealing that high doses of the drug tripled the risk of heart attack. The FDA estimated it caused 27,785 heart attacks. Ultimately Merck settled the lawsuit for $4.85 billion and later paid a $950 million fine to the DOJ. After the civil settlement in 2005, Merck CEO Raymond Gilmartin took an early retirement and a job teaching a course titled "Building and Sustaining Successful Enterprises" at the Harvard Business School.[202]

In the popular imagination, the criminal drug trade is synonymous with cartels and cocaine. And yet, under the cloak of scientific respectability, much organized crime has been conducted by certain American pharmaceutical companies. For the last fifty years, the world's closest observer and chronicler of medical and pharmaceutical malfeasance has been the psychiatrist and author Dr. Peter Breggin.

CHAPTER: 37

The Conscience of Psychiatry

A tragic feature of the human condition is that it takes many years to acquire great experience and wisdom, and as individuals reach an advanced age, their faculties decline and they retire from public affairs. During dangerous and confusing times, when wisdom is needed the most, exceptionally wise individuals may no longer be at hand to make sense of what is going on and to enlighten the rest of us. Not so with Peter R. Breggin, MD. When COVID arrived, the practicing psychiatrist and author was just shy of 84, but his capacity for observation and research—which he'd steadily honed during fifty years of closely watching the medical and pharmaceutical industries—hadn't diminished at all.

After graduating from Harvard College, he attended the Case Western Reserve Medical School. He then completed his postgraduate training in psychiatry, which included a residency at Harvard's Massachusetts Mental Health Center with a teaching fellowship at Harvard Medical School. Over the years he worked at the National Institute of Mental Health (NIMH) and held teaching posts at various universities. In the sixties he perceived the growing influence of the pharmaceutical industry in the mental health profession, especially in institutional settings. He didn't want to be a part of it, so he went into private practice in 1968.

In 1971 he was distressed by the resurgence of interest in performing lobotomies and psychosurgery of normal brain tissue for psychiatric or behavioral purposes. Proponents of the crude surgery believed it might be useful for correcting violent perpetrators such as the leaders of the Detroit Riots of 1967. A prominent advocate of experimental psychosurgery was the

Harvard Professor of Surgery, William Sweet. In the year 1970, Professor Sweet's colleagues Professors Vernon Mark and Frank Ervin published the book *Violence and the Brain,* in which they documented four case studies of psychosurgery using electrodes to destroy parts of their patients' brains. Though their book presented their work in a promising light, in fact the four patients experienced no long-term positive results, and two of them were badly impaired by the procedures.[203]

Violence and the Brain was published on the heels of *Physical Control of the Mind: Toward a Psychocivilized Society,* by the Spanish-born Yale neuroscientist José Delgado. Professor Delgado's book detailed how animals (including a Spanish fighting bull) and humans can be controlled through electrical stimulation of the brain. Dr. Breggin was horrified to learn that these eminent academic authorities were conducting aggressive experimental procedures and surgeries that had no basis in science. Their knowledge of the brain was extremely limited, but out of hubris, they presumed to improve a patient's personality and conduct by electrifying or destroying parts of his frontal lobe—the seat of our high order awareness of love, good, evil, abstract thinking, and planning.

At around this same time, Dr. Breggin learned that researchers at the University of Mississippi in Jackson were performing experimental psychosurgery on black children. And so, in the year 1972, he launched a full-blown campaign against the practice. Working with Congress, the Black Caucus, and conservative US Senators, he succeeded in putting a halt to experimental lobotomies.

Years later, in conversations with his wife, Ginger, he realized that the proponents of psychosurgery weren't just arrogant Dr. Frankensteins playing God with individual subjects, but were taking the first steps toward transhumanism. *Controlling and transforming humanity by medical means.* This was the ambition. The CIA's MK Ultra project of experimenting with LSD and other mind-altering drugs was animated by the same ambition.

Another major research project of Dr. Breggin was analyzing the role of German doctors and psychiatrists in formulating the Nazi eugenics program, and ultimately the Holocaust itself. He discovered that many of the regime's worst practices were proposed by the German medical establishment, which had been among the most respected in the world prior to 1933.

Because of their authority (real and perceived) about matters of life and death, medical doctors were in an ideal position to exert influence in formulating government policy. What began as apparently rational, scientific assessments of humans could lead to the piles of dead and emaciated bodies that Dr. Breggin had first seen in a Movietone newsreel when he was nine years old. Until that moment he hadn't given much thought to his Jewish ancestry, so it was an exceptionally eye-opening experience. This, it seemed, was what humans wielding unchecked power were capable of doing. His research of this darkest chapter in human history gave him a lifelong appreciation for the Nuremberg Code's regulation of medical practice.

Dr. Breggin's protest of lobotomies was the beginning of his career as "the conscience of psychiatry," as he later came to be known. He also protested electroshock therapy, which culminated in a showdown at the 1972 meeting of the American Psychiatric Association, at which proponents of electroshock therapy insulted and tried to intimidate him. He sued them for defamation and won a significant settlement.

For decades, Dr. Breggin has been a leading critic of the biochemical model of mental health. This model rests on the fallacious claim that mental suffering is caused by imbalanced or deficient brain chemistry that can (and should) be corrected by administering psychiatric medications and other biological interventions including electroshock. In 1983, he published his first critical book on the subject, *Psychiatric Drugs: Hazards to the Brain*. His 1994 book *Talking Back to Prozac* was a bestseller about Eli Lilly's blockbuster Selective Serotonin Reuptake Inhibitor.

He started researching the drug after seeing reports of patients experiencing dangerous adverse reactions, including violence, suicide, and murder. His inquiry led him to conclude that Prozac's efficacy had been grossly overstated, and that whatever benefit it may have conferred was greatly outweighed by its dangerous side effects. Eleven years later, the FDA partly recognized the validity of his findings by requiring black-box warnings on SSRIs for suicidality and overstimulation causing hostility, agitation, insomnia, and mania.

Ultimately Dr. Breggin realized that the Psycho-Pharmaceutical Complex (as he referred to it in a 2008 book) was far more interested in making money than helping patients. The Complex advanced its agenda by claiming to possess the "science" of the brain, thereby indoctrinating naive psychiatrists and their patients with the biochemical model of mental health. The result was that many therapists adopted the practice of prescribing psychiatric drugs at their patients' first session, without getting to know them. This practice ignored the fact that mental suffering usually arises from real things in the patient's life that cause him distress and that can be corrected with psychotherapy, education, and encouragement.

From more than fifty years of study, observation, and clinical practice, Dr. Breggin had become as intimately acquainted with human nature as anyone in history. One of the most important lessons he'd learned is that, at any given time, humanity contains a certain number of highly predatory individuals. Back when we lived in hunter-gatherer tribes, such predators—with their exceptional energy, drive, and aggression—probably provided useful leadership. However, after we organized into large social and political units, the predators among us became power-hungry tyrants and dictators who pursued their ambitions at the expense of humanity, often causing great destruction and suffering. All of history, from Biblical times and the Roman domination of Europe, up to 20th-century totalitarian dictatorships, has been shaped by predatory individuals building empires.

When COVID-19 arrived, Dr. Breggin and his wife, Ginger, were unusually prepared to take on what they would call the "Global Predators," having researched twenty books and sixty scientific papers examining corruption in the scientific, medical, and psychiatric establishments, and how their corruption extended into federal institutions such as the NIH, FDA, and DOJ. Additionally, Dr. Breggin had been approved to testify as a medical and psychiatric expert in US federal and state courts, as well as Canadian courts, over 100 times.

Equipped with this experience and expertise, the Breggins quickly recognized that SARS-CoV-2 was made in a lab and that the international policy response was being dictated by the same network of predatory special interests that created the monster. They documented each day's events and communicated with other researchers who were also observing and analyzing the story as it unfolded in real time.

In August 2020 they began writing their book *COVID-19 and the Global Predators: We Are the Prey,* a masterpiece of exhaustive, investigative reporting about every aspect of the man-made disaster. The book meticulously dissects the ten-year plan leading up to COVID-19 involving Bill Gates, Klaus Schwab of the World Economic Forum, Anthony Fauci, Tedros Ghebreyesus of the World Health Organization, and many other powerful forces, including the pharmaceutical and banking industries.

In the autumn of 2020, Dr. Breggin contacted Dr. McCullough, who shared his experience and research in early treatment, and wrote an introduction to the Breggins' book. Around the time they finished writing it, a nonfiction book was published in the United States about one of the great apex predators of the American pharmaceutical industry.

CHAPTER 38

Empire of Pain

About seventeen years before COVID-19 arrived, I started hearing stories about people in my extended social network getting hooked on OxyContin and causing immense damage to themselves and their families. How, I wondered, had such a destructive substance been marketed and sold in the United States for so long, and what did it say about our medical profession and regulatory agencies?

In 2017, I read a fascinating piece in the *New Yorker* titled "The Family that Built an Empire of Pain."[204] It was the story of the Sackler family of New York and their company, Purdue Pharma, which reportedly made $35 billion selling OxyContin. I'd already sensed that the OxyContin story was an indictment of the US Bio-Pharmaceutical Complex, but I was still stunned by the scope of the corruption described in this article.

From 1996 to 2001, Purdue Pharma used a network of academics, doctors, lobbyists, publicists, regulatory agency friends, medical associations, and an army of sales reps to push its opioid. At the heart of this operation was the systematic denial or downplaying of the drug's (long understood) addiction risk. Consequently, millions of Americans became addicted to OxyContin with catastrophic consequences. For many, the substance was a gateway drug to more dangerous opioids such as fentanyl. The OxyContin story was not an aberration, but the culmination of Arthur Sackler's extraordinary work in developing Big Pharma's bag of tricks.

"Most of the questionable practices that propelled the pharmaceutical industry into the scourge it is today can be attributed to Arthur Sackler," explained Allen Frances, former chair of psychiatry at Duke University

Medical Center, to the *New Yorker*. A few years before Sackler acquired the fledgling drug company Purdue Frederick in 1952, he acquired controlling interest in the advertising firm William Douglas McAdams, where he spearheaded many of Pfizer's top PR campaigns. Sackler pioneered the practice of placing "informational" pieces about pharmaceuticals in the *Journal of the American Medical Association*. He was also the first to suggest that sales reps give samples of their products to doctors.

When Pfizer launched its Atarax tranquilizer in 1957, Sackler promoted it with extraordinary aplomb. The central pillar of his campaign was a thirteen-minute "public service presentation" titled "The Relaxed Wife" that ended with the exhortation, "If you have tension problems, discuss them with your doctor." Soon the phrase "discuss with your doctor" became the ubiquitous drug marketing formula that it remains today.[205]

Sackler also spearheaded the campaign for Hoffman La-Roche's valium with the bold claim that it was the right choice for *everyone*, even if they didn't suffer from nervous tension. Referring to naturally calm people, he recommended to doctors: "For this kind of patient—with no demonstrable pathology—consider the usefulness of valium." Sackler's "public service presentations" were so successful in promoting tranquilizer use among American housewives that the Rolling Stones would sing of the common habit in their 1966 hit "Mother's Little Helper."

When Arthur died in 1987, he bequeathed his playbook to his brothers, and in 1995 their company, renamed Purdue Pharma, received FDA-approval for OxyContin. As the *New Yorker* reported the process:

> Purdue had conducted no clinical studies on how addictive or prone to abuse the drug might be. But the F.D.A., in an unusual step, approved a package insert for OxyContin which announced that the drug was safer than rival painkillers, because the patented delayed-absorption mechanism "is believed to reduce the abuse liability." . . . The F.D.A. examiner who oversaw the process, Dr. Curtis Wright, left the agency shortly afterward. Within two years, he had taken a job at Purdue.[206]

The OxyContin story was a shocking example of how the representation of a phenomenon—in this case a dangerously addictive substance—can be manipulated on a massive scale. For years, US institutions, agencies, and the mainstream media somehow failed to notice what OxyContin was doing to American society. It was only in 2017 that the *New Yorker* published Patrick Radden Keefe's report. In that same year, the U.S. Department of Health and Human Services declared the Opioid Epidemic a public health emergency and estimated that 70,000 Americans would die of opioid overdose in that year alone. According to the CDC, between 1999 and 2019, nearly 841,000 people died from a drug overdose. More than 70 percent of overdose deaths were caused by an opioid prescription, heroin or fentanyl. During the first year of the COVID-19 pandemic, that number rose to 100,306.[207] Most of the victims were younger than fifty-five years old.

On April 17, 2021, Patrick Radden Keefe's comprehensive book on the story, *Empire of Pain*, was published to great critical acclaim and popularity. Major newspapers and television pundits gave the book a glowing review, but none remarked that the corruption it revealed within the US Bio-Pharmaceutical Complex was relevant to the COVID-19 pandemic. Three weeks after *Empire of Pain* was published, Dr. Peter McCullough was interviewed on *Tucker Carlson Today*.

CHAPTER 39

The Philosopher

I watched a recording of the interview. When Mr. Carlson asked him to explain what was going on and McCullough replied, "That's going to be the goal of investigative journalists to figure out," I sensed he knew a lot more than he was ready to discuss on television. Clearly, he wanted to stay in his lane as a doctor and avoid speculating about criminal intent and motive. I called HeartPlace, got his e-mail address, and invited him to an interview.

Two days later we met at the Deep Ellum district studio of Justin Malone—a documentary filmmaker whose 2020 film *Uncle Tom: An Oral History of the American Black Conservative* had reached and touched an unexpectedly large audience, despite being ignored by the legacy media. The prominent black men and women who told their stories in the film offered an unorthodox—some might even say heretical—view of race in America. Researching the film awakened Justin to how much public perception is shaped by propaganda, and he graciously offered to film the two-hour interview.

The first thing that impressed me about Dr. McCullough was his trusting nature. Though I did a pretty good job of introducing myself by e-mail, he still had no means of knowing what he was getting himself into. At the appointed time, I stood on the sidewalk outside of the studio. A large Dodge pickup truck rounded the corner, and the slender, slightly built man got out wearing a suit and tie. The truck's window rolled down, and I saw a pretty brunette woman sitting behind the wheel. She warmly greeted me with a bright smile as Dr. McCullough introduced her as his wife, Maha.

We went into the studio, where he sat down in the interview chair and told us about his ongoing efforts to treat COVID-19, and the strange resistance he'd encountered. Instantly I sensed that what animated him was a burning desire to learn and tell the truth. Such people are extremely rare. Plato described this kind of character as a philosopher—literally a "lover of wisdom." He seeks to understand reality and will pursue it at all costs, including his personal security and even his life.

Throughout history, great philosophers have found themselves at odds with the ruling class, because telling the truth tends to challenge powerful interests. Notable historic examples were Socrates, Giordano Bruno, and Galileo. All three were tried and convicted of impiety or heresy. The first two were put to death; Galileo spent the last nine years of his life under house arrest. Another notable scientist prosecuted for heresy was Michael Servetus, the first European to correctly describe the function of pulmonary circulation. He was burned at the stake in Geneva in 1553.

My main objective with the interview was to evaluate the official policy response to COVID-19 by the medical standards that McCullough had always known. He burned red hot in the interview. Every sentenced flowed with no pauses. He took no prisoners and left no doubt. As I'd suspected, most COVID policy was a radical departure from long-standing principles of caring for patients, and a glaring breach in the fiduciary relationship between physician and patient, and hospital to community.

Especially baffling was the therapeutic nihilism that permeated the medical establishment and was blindly accepted by the press. Soon after SARS-CoV-2 arrived, the federal health agencies told the medical profession that *nothing* could be done to treat the disease. The press embedded this message in every narrative. All we could do was stay at home and wait for a vaccine.

At the interview's conclusion, I escorted McCullough out and then went back into the studio for a meeting with the production crew. They were still in a state of shock by what they'd heard.

"I don't think I've ever heard such a terrible story," Justin said. "I think *evil* is the only way to describe it?" We reviewed the tape, and Justin marveled that he wouldn't need to make a single cut to the interview.

"He speaks with such mastery and intensity," Justin remarked. "There's not an ounce of fat on it. I think we should post the uncut interview with no copyright claims. The world should see this as soon as possible." The next day I posted it on YouTube, and it was taken down a few hours later. I then posted it on Vimeo, as did Justin, and soon the video went viral on several platforms, especially Rumble.[208]

CHAPTER 40

Graduating into Eternity

One of the persons who saw the video was Jodi Carroll. She watched it with great interest and in particular noted McCullough's remarks about the usefulness of ivermectin and corticosteroids in treating hospitalized patients. A month later, her 74-year-old mother, Carolyn, fell ill with COVID-19 and was admitted to Baylor, Scott & White hospital in College Station, Texas. Jodi, who had been visiting friends in Colorado, was alarmed and immediately sent an e-mail to Dr. McCullough through the contact form on the HeartPlace website. She then packed and got on the road to College Station, her mind racing with fearful questions as she reflected on the terror her mother must have been experiencing.

Jodi checked into a hotel directly across the highway from the hospital where her mother lay, fighting for her life. The little hotel room soon became her prayer chapel. Every few hours she ventured across the road to glean whatever information she could. Just as she'd feared, the hospital's staff refused to administer ivermectin to her mother. Nor did they administer the methylprednisolone corticosteroid in the FLCCC protocol, nor did they administer the full-dose aspirin and lovenox anticoagulants in the McCullough protocol. According to Carolyn's chart, the hospital initially administered 10 mg of dexamethasone per day. This was, in the words of Dr. Paul Marik, "the wrong drug, the wrong dose, and the wrong duration of Rx." As for the anticoagulants: the hospital administered 81 mg of aspirin once daily and 40 mg of lovenox twice daily. The McCullough protocol recommended 325 mg of aspirin per day and 80 mg of lovenox twice daily.

Jodi showed my interview with Dr. McCullough to her father, Lee, who had Medical Power of Attorney, and he too found the doctor's arguments persuasive. Together, he and Jodi repeatedly requested that hospital staff administer ivermectin and the FLCCC protocol to her mother, but each time the staff refused. Carolyn Carroll did not get better in the days following her admission. On the contrary, she declined and appeared to be headed for the ventilator. One day, as Jodi sat in her hotel room, immersed in prayer, beseeching God for help, her cell phone rang.

"Jodi, this is Dr. Peter McCullough, returning your message. Tell me what is going on with your mother."

After a brief explanation, he said he would do his best to help but cautioned that he no had means of compelling the hospital to administer the therapy. His heart ached for the woman, because he knew well the immense power she was up against. He was especially concerned about blood clotting in the lungs, so he told Jodi to press hard for the full dose (325 mg per day) of aspirin and a full dose (80 mg twice daily) of lovenox. The hospital insisted on maintaining only 81 mg per day of aspirin. It made a partial concession by increasing her lovenox dose from 40 mg twice daily to 60 mg twice daily, though this was still far below the 80 mg twice daily that McCullough recommended. To provide supporting documentation for her family's request, Jodi printed out copies of the relevant academic medical papers and gave them to her mother's medical team.

"The doses you are asking for could cause internal bleeding," a doctor said to Jodi. "Blood could start rushing out of her colon. You wouldn't want to be responsible for that, would you?" She reported this remark to McCullough, who replied that the risk of internal bleeding should be assessed in comparison to the risk of the patient dying from blood clots in her lungs.

"I guarantee you that other patients in that hospital are receiving the exact doses of the same drugs you're asking for, but for some mysterious reason, the hospital refuses to give the same regimen to COVID-19 patients."

So began Jodi's long and frustrating effort to induce the hospital to give her dying mother the drugs that McCullough recommended. After all of her polite pleas were ignored, she felt she was left with no choice but to increase the pressure due to the obvious urgency of the situation. Her brother, Clover, and sister-in-law, Rachel, sought an ethics consult with the hospital's designated patient advocate. They didn't hear back from the woman for two weeks. When they finally did, she offered to obtain additional blankets and pillows for the patient, but she seemed unable to provide anything else. Later, Jodi discovered that an ethics executive had responded to her family's request by calling the patient herself—while she lay unconscious on a ventilator in the ICU.

McCullough put Jodi in touch with an attorney named Todd Callender, who was working with America's Frontline Doctors. Mr. Callender submitted a letter to the hospital, asking that it administer the McCullough Protocol pursuant to federal and state "Right-to-Try" laws, but his appeal was rejected.

The Carrolls also appealed to a member of their extended family. The retired medical malpractice trial attorney happened to be a benefactor and former board member of the hospital. He'd long had a close relationship with Carolyn, who had provided him with daily care and assistance during his convalescence from a recent illness. He passed along the request to the chief medical officer, who in turn contacted Jodi's father. Jodi was with him at his desk with his phone on speaker mode when the hospital executive called.

Again, Jodi politely requested the McCullough Protocol for her mother, stating the precise medicines and doses. The chief medical officer asked her a series of questions about the drugs and their rationale and then promised to look into it. In fact, he never got back to Jodi or her father. Later her father received a call from their family member—the hospital board member and benefactor—who exclaimed, "Tell Jodi to leave the hospital's doctors alone and let them do their job!" And yet, as far as Jodi could see, the staff was doing little more than watch Carolyn slip away while refusing to give her FDA-approved, repurposed drugs that *could* help her.

Jodi's brother and his wife spoke with Texas Senator Lois Kolkhorst, who was especially sympathetic to their cause from her own family's recent experience with COVID-19. The senator contacted the hospital to try to advocate for Mrs. Carroll—again to no avail. At the 85th District Court in Bryan, Texas, Jodi and her father filed for a mandatory injunction compelling the hospital to administer the McCullough Protocol to Carolyn Carroll. The court ordered Baylor to administer the drugs, but the hospital's attorneys appealed and were granted another hearing, where they argued that the Hippocratic Oath compelled them first and foremost "to do no harm." The hospital's expert witness, Dr. Seth Sullivan, claimed that the proposed drugs were "unproven and possibly dangerous" and therefore could not be lawfully administered to the patient. The judge apparently found this argument persuasive and overturned the court's initial order. Jodi and her family were devastated.

During a recess, Jodi received a call from Dr. McCullough, who mentioned in passing that he too was in a legal contest with Baylor, Scott & White, which was suing him. This news struck Jodi with grief upon grief, as she assumed the lawsuit was retribution for giving her the medical advice that was the basis of her legal action against Baylor.

"Don't worry," he said. "My troubles with Baylor go way back and have nothing to do with you."

Seeing that Jodi was at the end of her tether, McCullough invited her to visit him and his wife in Dallas. I joined the three of them for dinner, and the next day I met Jodi for coffee. As she told me about her struggle, I sensed she was in a state of acute distress that was perhaps akin to PTSD. Her encounter with Baylor's hired-gun attorneys and their expert witness had been especially bruising.

"Why would they go to such lengths to avoid giving my mom ivermectin and the other medications?" She asked. "The doctors admit that her condition is worsening and that she's probably not going to make it. Why won't they let her try something, anything that could help her?"

"God, I wish I knew," I said. We concluded our talk, and then I walked her to her car. "I'm so sorry, Jodi," I said. "I wish there was something I could do to help."

"Thank you," she said and got in her car and drove away. The next day I got a text notification and looked at my phone and saw it was from her.

Mom graduated into eternity today at 2:50 p.m.

Jodi was exactly my age, and her mother was exactly my mother's age. I imagined the despair and rage I would feel if I'd been in her shoes, and I wondered what I would do. Her helplessness as she watched her mom slip away—while the hospital steadfastly refused to give her something that could have helped her—must have been akin to that of a woman being physically violated.

At McCullough's urging, Jodi requested an autopsy on her mother. As he'd anticipated, the lungs were filled with blood clots. This was the final death sequence he had endeavored to prevent when he and his colleagues formulated the sequential multidrug treatment. Administering it to Mrs. Carroll could have halted the disease's progression and enabled her to enjoy some more good years.

CHAPTER 41

The Stripping

In his Separation Agreement with Baylor, Scott & White Health, dated February 24, 2021, McCullough agreed not to state that he is employed by or affiliated with the institution after departing his employment. McCullough honored the Agreement. In none of the countless interviews he gave in the spring did he state that he was employed or affiliated with Baylor Scott & White. However, in a few instances *after* he gave the interview, the postproduction crew Googled his name for his professional titles, which they put on banner bars that appeared when the interview was posted or broadcast. In the spring of 2021, a top search result for his name was a cardiology conference sponsored by Harvard at which he gave a lecture. The lecturer's bio stated he was the "Vice Chief of Internal Medicine at Baylor University Medical Center," and so this was the title that made some of the banners.

In the late spring, the topic of McCullough's media discussions increasingly shifted to the subject of vaccine safety data, which he found extremely alarming. According to the CDC, 6,207 deaths of people who'd received the COVID-19 vaccine were reported to the Vaccine Adverse Events Reporting System (VAERS) up to July 26, 2021. This was a staggering number. By comparison, the 1976 Swine Flu mass vaccination program was shut down after about 25 deaths and 550 cases of Guillain-Barré syndrome were reported. McCullough pointed this out in his media interviews, apparently to the consternation of Baylor administrators.

No one from Baylor contacted him by e-mail or phone to express concern that this was happening and potentially damaging the institution's reputation. Nor did McCullough ever receive a cease-and-desist letter. Baylor said nothing until July 28, 2021, at which point it filed a lawsuit, claiming that the false representations of McCullough's affiliation were "likely to cause Plaintiffs irreparable reputational and business harm that is incapable of remedy by money damages alone. This is particularly true in the middle of a global pandemic." As a result, the plaintiff "seeks monetary relief over $1,000,000 from Defendant and nonmonetary relief."

McCullough learned he was being sued when his wife, Maha, called him at work and told him that the notice had just arrived in the mail. This obliged him to hire a local litigator to represent him, which began a lengthy interchange with a Dallas plaintiff's attorney whom Baylor had retained to litigate the suit. Soon the legal bills started rolling in for up to $40,000 per month, far in excess of his HeartPlace salary. The Baylor attorney said she wanted access to all of McCullough's e-mails, and when he refused, she told his lawyer that unless he provided access to his e-mails, she would file a motion to compel discovery. Baylor was barking up the wrong tree, because not once after his official termination on February 24 did McCullough mention in an e-mail that he still worked for Baylor or that his opinions were endorsed by Baylor staff.

The case went into mediation, and McCullough said he would agree to Baylor's requests for how he presented himself and his CV. In return he asked Baylor to provide a written statement that it had incurred no damages from his media activities. His attorney advised him that the plaintiff would never agree to this condition, to which McCullough replied that if they refused to agree to it, they would have to specify what damages they had incurred. After McCullough submitted this request, Baylor didn't respond and then fell silent for long periods of time—dragging out the litigation with an apparent goal of never resolving the matter.

The press picked up the story of Baylor's lawsuit, starting with a *Dallas Morning News* report on July 29 under the headline "Baylor Health sues vaccine skeptic and demands Dallas doctor stop using its name."[209] Shortly after word got out that McCullough was being sued, he began receiving letters from institutions, terminating his affiliation. The first to arrive was from the dean of Texas A&M University, stripping him of his professorship. A month letter he received a second letter from the university general counsel, threatening to sue him if his former professorial title appeared again in any media posting. Then he received a letter from the American College of Physicians, warning him that the college was monitoring his media activities and that his statements could bring him into conflict with the ACP. McCullough responded by voluntarily delisting himself from the college after being a dues-paying member for thirty-five years.

Next was a letter from Texas Christian University, stripping him of his professorship. Then a letter from the University of North Texas Health Science Center School of Medicine, kicking him off the faculty. Then came a letter from the *Journal of Cardiorenal Medicine*, of which he'd been the editor in chief for five years, terminating his editorship. Then a notice from the American College of Cardiology, terminating him from a major research paper he'd worked on with the institution for two years. Then one from the National Institutes of Health, ejecting him from an external advisory panel on a major epidemiological study. Then notices from pharmaceutical companies, ending his advisory positions on clinical trials.

The Internet was yet another front in the war declared on his character and professional existence. An ad hominem assault with epithets such as "quack, discredited, debunked, liar, and grifter" was conducted in a barrage of tweets, comments, and posts with no accountable sources or citations, and without substantiation from any medical authority of equal academic stature to McCullough. He, on the other hand, was censored, and his social media posts were labeled as misinformation by self-appointed (and usually anonymous) "fact checkers" who cited no studies to support their assertions.

The most bizarre moment in his stripping was the notice that the Cardiorenal Society of America, of which he'd been president for five years, was dissolving. *What on Earth?* McCullough thought. *An entire medical society dissolves because its long-standing president advocates treating COVID-19 and cautions about vaccine safety?* Truly the world had lost its mind.

The dismantling of McCullough's academic career resulted in considerable financial losses, but these weren't what grieved him. He thought of the decades of study, work, lectures, publications, and grant proposals he'd poured his heart and soul into. All the relationships he'd built on a foundation of mutual respect; all the recognition and honor he'd received—all was nullified by a series of icy letters, leaving him with a heap of broken memories of his career.

Impenetrably strange was his ostracism's total lack of ceremony. This struck him as being akin to his oldest friends simply ghosting him. Sometimes, half-joking, he thought it would have been preferable if Baylor had hosted a going-away party for him in which the institution declared that he'd lost his mind but still wished him well. At least that would have given him some sense of ending.

CHAPTER 42

Other Assassinations

At the same time McCullough was being stripped, his colleague and kindred spirit, Dr. Paul Marik, experienced a similar fate. On October 15, 2021, his hospital's administration circulated a memo to the entire healthcare system stating that its doctors were authorized to administer remdesivir to COVID-19 patients, but not ivermectin or a host of other repurposed drugs. As Dr. Marik read the memo, he marveled at the sheer perfidy of it. Especially grotesque was the inclusion of "Ascorbic acid" (vitamin C) on the list of banned substances.

The administration issued this directive at a time when seven COVID-19 patients were in the ICU, desperately in need of Dr. Marik's care. He, in turn, desperately wanted to treat them with the drug regimen that he knew would give them a good chance of recovery. At a US Senate hearing three months later, he recounted his helplessness:

> This system was effectively preventing me from treating my patients according to my best clinical judgment. . . . As a clinician for the first time in my entire career, I could not be a doctor. I could not treat patients. I had seven COVID patients [he holds up his hands showing seven digits] including a 31-year-old woman. I was not allowed to treat these people. I had to stand by idly [he clenches and raises his fists with anguish and begins to weep]. I had to stand by idly, watching these people die.
>
> I then tried to sue the system, so then they did something called peer sham review. It is a disgusting and evil concept. They then accused me of seven most outrageous crimes . . . and [claimed] that I was such a severe threat

> to the safety of patients, they immediately suspended my hospital privileges because I posed such a threat to these patients—ignoring the fact that under my care, mortality was 50 percent less than it was under my colleagues. I then went on to this sham peer review. I went to a Kangaroo Court, where they continued this, and the end result was that I lost my hospital privilege and was reported to the National Practitioner Databank. So here I was standing up for my patients' rights, and this hospital, this evil hospital, ended my medical career.[210]

Drs. Pierre Kory and Umberto Meduri at the FLCCC were also stripped of their positions. In France, Professor Didier Raoult was subjected to all manner of harassment. Countless other doctors with hospital affiliations such as Ryan Cole, Simone Gold, Brooke Miller, and Mary Talley Bowden were fired or lost their hospital privileges. Independent doctors such as Vladimir Zelenko, Ivette Lozano, George Fareed, and Richard Urso were harassed by state medical boards and vilified in the media—all for the crime of advocating the early treatment of COVID-19 with drugs that had long been FDA-approved for other conditions. Along with Dr. McCullough, they are the heroes of this story.

Another man who has boldly challenged the suppression of early treatment is Robert F. Kennedy Jr. McCullough twice went on his *Defender* podcast and was a key interviewee for his book *The Real Anthony Fauci*, published in November 2021. Despite suffering from spasmodic dysphonia, RFK Jr. has, for the last two years, been one of America's strongest voices in defending our Constitutional Republic from public health officials and politicians wielding emergency power. In a series of rousing speeches reminiscent of his father's famous University of Capetown address in 1966, RFK Jr. has articulated why we should never allow our Constitution to be compromised by fallible men who promise to keep us safe.

For decades, he worked as an environmental attorney to protect the natural world from corporate industrial polluters. He has been especially troubled by the contamination of our waterways with hazardous waste, including mercury. In 2005 he became concerned about a mercury compound used as a preservative in childhood vaccines, which prompted him to conduct a thorough investigation of vaccine safety in general. Though some of his conclusions may be debatable, the issue of vaccine safety is a debate worth having. The 1986 National Childhood Vaccine Injury Act granted vaccine manufacturers immunity from all civil and criminal liability for injuries or deaths caused by their products. In the absence of legal liability, the only thing likely to regulate their conduct is scrutiny by a public figure like RFK Jr.

His criticism of vaccine makers has not been met with debate, but with vitriolic, ad hominem attacks and accusations of being a conspiracy theorist. Implied in these attacks is that no one should dare even question the safety of vaccines. Every statement in his book *The Real Anthony Fauci* is documented with a citation of primary sources including federal agency documents, peer-reviewed medical literature, and public records. Any reader can easily evaluate these sources. The Kindle edition features hyperlinks to the documents. Nevertheless, not a single newspaper reviewed the book, and not a single network or cable broadcaster interviewed the author. Despite this media blackout, *The Real Anthony Fauci* has sold over a million copies. RFK Jr. donated the proceeds to his Children's Health Defense charity.

CHAPTER 43

The Joe Rogan Experience

Around the end of November 2021, I worried that Dr. McCullough's ceaseless efforts to tell the truth were destined to be in vain because of the vast power he was up against. Every day was a fresh battle against censorship. The forces who'd campaigned against early treatment were now doubling down on their effort to "get a needle in every arm" in spite of a plethora of documented myocarditis cases following the injections, most notably among young, athletic males. This was especially disturbing, given that this cohort faced zero risk from COVID-19.

With December's arrival, I felt downright despondent. And then, just as it seemed that all might be lost, McCullough told me that Joe Rogan had invited him for an interview. This was tremendous news. For years I'd been a Joe Rogan fan. Scarcely a day passed without listening to one of his conversations. His courage to talk about anything with anyone, and his calm, steady voice of reason had become a comforting presence in my life.

A few months earlier, Rogan had his own brush with Bio-Pharmaceutical propagandists. It began on September 1, when he announced he had COVID-19 and was treating it with ivermectin prescribed by a doctor. CNN's entire roster of hosts mentioned it in their shows, all proclaiming that Rogan was taking a "horse dewormer." About six weeks after this incident, Rogan hosted CNN's chief medical correspondent, Sanjay Gupta, who at one point in the interview said, "By the way, I'm glad you're better."[211]

"Thank you. You're probably the only one at CNN who's glad. . . . The rest of them are all lying about me taking horse medication," Rogan replied.

"That bothered you," said Gupta.

"It should bother you too," Rogan answered. "They're lying at your network about people taking human drugs versus veterinary drugs."

"Calling it a horse dewormer is not the most flattering thing."

"It's a lie," Rogan shot back. "It's a lie on a news network . . . and it's a lie that they're conscious of. It's not a mistake. They're unfavorably framing it as veterinary medicine. . . . They know it's a human drug and they lied. It's defamatory."

McCullough's interview at Rogan's Austin, Texas, studio took place on December 8. He brought his laptop so that he could refer to slides of the peer-reviewed studies he could cite for each data point. Rogan was intensely curious about the studies and looked them up on his own computer as they went along. Never once did he interrupt the presentation of data, and McCullough marveled that his millions of young listeners would apparently hold still for a three-hour academic medical conversation. Somehow, Rogan had shattered the mold of what the legacy media considered a viable format and in so doing had attracted a listenership that dwarfed theirs. While Anderson Cooper might draw 1 million viewers per show and Tucker Carlson 3 million, Rogan regularly drew 11 million.

The day after McCullough's interview, I met his media assistant. She told me it had gone well and would soon be posted on Spotify. Five minutes later her cell phone rang. It was a friend of hers in Austin who'd facilitated the interview with the *Joe Rogan Experience*. He had bad news.

"Spotify is pressuring Joe to axe it," he said. Joe was valiantly pushing back, but the terms of his contract with Spotify did not give him full discretion. *Come on*, I thought. *Surely Joe Rogan is too big and revered for his dedication to free speech to be censored.* I got on the phone with the man in Austin.

"How do you think this is going to play out?" I asked.

"It seems they really don't want Joe to post it."

"What would happen if he just put his foot down?"

"They paid him a lot of money for his contract," he said and stated a very large sum. *Money*, I thought. An image of the weird, slightly sinister Master of Ceremonies (played by Joel Grey) in Bob Fosse's film *Cabaret* flashed through my mind, and Grey's duet of "Money Makes the World Round" with Liza Minnelli. *Money makes the world go round / Of that we both are sure.* Suddenly I felt annoyed.

"Well, maybe money's not everything," I said. The young man repeated the sum that Spotify is paying Joe Rogan. In his mind, it seemed there was no arguing with the number. I told him the apocryphal story of Winston Churchill asking a smug lady if she would sleep with him for a million pounds.

"Maybe," she replied.

"Would you for five pounds?" he asked.

"Winston, what do you take me for?"

"We've already established that, now we're just negotiating the price."

The young man on the phone didn't find this anecdote funny.

"Well, I hope Joe will fight for it," I said. "Someday he'll be glad that he did."

Mr. Rogan did fight for it, and when he posted the interview a few days later, it drew 40 million views in its first week.[212] Multiple commentators accused McCullough of "peddling false claims" and called for the podcast to be removed. A few weeks later, Dr. Robert Malone amplified the kerfuffle in a Joe Rogan interview in which he corroborated many of McCullough's statements.[213] This proved to be too much for the Canadian musicians Neil Young and Joni Mitchell. The former had once sung the imperative to "Keep on Rocking in the Free World." The latter had once exhorted farmers to put away their DDT. Now they were outraged that highly credentialed doctors were allowed to speak on a public platform about the risks of experimental medical products that many were being forced to take in order to retain their jobs. The manufacturers were making billions while bearing no liability for injuries or deaths caused by their products. Did Mr. Young and Ms. Mitchell really believe that no one should be allowed to question this situation?

It is irrational and morally untenable to claim that Drs. McCullough and Malone have no right to question something unless they already possess perfect understanding of what mankind is facing. Both men have formed their opinions based on rapidly emerging data about a complex reality, much of which lies at the limit of human understanding. Instead of blindly accepting official representations of this reality, they have critically examined and questioned these representations. This is how *all* serious inquiry has been conducted for thousands of years. To claim that Drs. McCullough and Malone aren't free to question and critically discuss medical orthodoxy is the equivalent of saying that no one is free to even think about it.

Of all the boneheaded developments the West has witnessed in recent years, the resurgence of censorship must be the stupidest. Those who commit it are subverting 350 years of wisdom accrued in the English-speaking world since John Milton gave his passionate defense of free speech before Parliament in 1644.[214] The US Founding Fathers understood the critical importance of free speech for maintaining a free republic, which is why they protected it with the First Amendment of the Constitution.

CHAPTER 44

Keeping Our Noses Clean

During his Joe Rogan interview, McCullough repeatedly mentioned what struck him as one of the most marvelous discoveries in the entire quest for early treatment—namely, that a simple nasal rinse using 1 percent povidone iodine in a saline solution could substantially reduce the SARS-CoV-2 viral load in the nose and sinuses. A few days before the interview, Professor I.M. Choudhury et al. at the Bangabandhu Sheikh Mujib Medical University in Bangladesh published the stunning results of a randomized clinical controlled trial in which 303 SARS-COV-2 positive patients performed the povidone iodine nasal rinse and gargle every four hours, while the control group of 303 SARS-COV-2 positive patients performed a nasal rinse and gargle of lukewarm water. On day 3, only 11.55 percent of the experimental group tested positive, while 96.04 percent of the control group tested positive. Only 10 percent of the experimental group progressed to hospitalization requiring oxygen support; 63 percent of the patients who didn't receive the therapy ultimately required hospitalization with oxygen support.[215]

The results of the Choudhury report were consistent with common sense and logic. Already in March of 2020, the *New England Journal of Medicine* published a report from Chinese researchers in Wuhan who'd found that in the initial phase of SARS-COV-2 infection, the virus replicates in the nose and sinuses before migrating down the respiratory tract and into the lungs.[216] Thus, it stood to reason that a nasal rinse—one that either destroyed the virus or inhibited it from binding to the cells—could reduce the viral load or clear it altogether. On July 23, 2020, Farrell et al. published a paper in *JAMA* titled "Benefits and Safety of Nasal Saline Irrigations in a Pandemic—Washing COVID-19 Away" in which they cited evidence that "hypertonic nasal saline, which facilitates mucociliary clearance, likely decreases viral burden through physical removal."[217] The authors also noted that adding diluted povidone iodine to the rinse could further reduce viral load and transmission.

Given that the therapy posed zero risk, it struck McCullough as a new summit of folly that no one in federal health agencies or in the mainstream media was talking about it. The more he investigated it, the more he became astonished that something as simple as rinsing the nasal passage could stop the disease progression before it really got started. Eventually his interest in the prospect led him to cross paths with an entrepreneur in Provo, Utah, who had been studying the benefits of nasal hygiene for twenty-two years.

Nathan Jones is the son of Dr. Alonzo Jones—a family physician in Plainview, Texas, who'd long had a passion for medical history. As every student of medical history knows, the greatest advances in human health were not achieved by drugs, but by public sanitation and personal hygiene. Dr. Jones was intimately familiar with the stories of Dr. Semmelweis in Vienna and Dr. John Snow, who lived and worked in London during the same period. Dr. Snow struggled mightily to persuade London's obtuse public health authorities that the cholera outbreaks that plagued the city were caused by sewage-contaminated drinking water. With remarkable thoroughness, he investigated a severe outbreak in 1854 and gathered evidence it was caused by a contaminated well in Soho. His proposition challenged the long-standing theory that cholera was caused by breathing a miasma of foul air. It took an enormous effort to persuade the Board of Health to remove the pump handle from the Broad Street well, and shortly thereafter the number of cholera cases precipitously dropped. And yet, in spite of his obvious success, Dr. Snow still encountered massive resistance from innumerable skeptics in the local medical and religious establishments. His final vindication came almost thirty years later when Dr. Robert Koch isolated the bacterium that causes cholera, which he called *Vibrio cholerae*. Koch demonstrated that this bacterium was present in contaminated drinking water, and not in airborne "vapors" or "miasma."

Oral hygiene—the removal of food particles from the teeth—originated thousands of years ago in India. Likewise, the practice of *jala neti*, or nasal washing, was presented in the Vedas—the foundational texts of the Hindu religion. Ear, nose, and throat infections were among the most common illnesses in Dr. Jones's family practice, which caused him to reflect on the Hindu ritual of daily nasal washing. Could it be that the ancient wise men who'd formulated it were really onto something? Could otitis media, asthma, sinusitis, and allergies be treated or even prevented by daily nasal hygiene?

In the mid-to-late 1990s, Dr. Jones read reports by Finnish research teams about the benefit of xylitol—a crystalline, water soluble "sugar alcohol" that cannot be metabolized by common bacteria that inhabit the mouth and inner ear. These bacteria consume the xylitol and are then impaired by it because they cannot digest it. The Finnish studies indicated that xylitol-infused chewing gum and syrup not only markedly reduced tooth decay, but also the incidence of ear infections by 42 percent in children. A Finnish study published in 1998 showed that xylitol reduced the adhesion of pathogenic bacteria to cultured nasal cells by 68 percent.[218]

Dr. Jones wondered if a saline-xylitol nasal spray could have a similar effect in promoting nasal hygiene. He formulated it with the objective of enhancing the normal cleansing function of the nasopharyngeal passages, which are constantly being challenged by air pollution, pathogens, and excessively dry air from modern HVAC systems. He reasoned that too much water and saline could wash away the nose's healthy mucosa in the same way that douching had been shown to do to the vagina. His objective was to create a nasal spray that matched the (slightly acidic) pH of the nose, with just the right solution of saline and xylitol for aiding the mucosa and cilia in clearing pathogenic and allergenic debris from the nose. Ultimately, he created a nasal spray consisting of an 11 percent solution of xylitol and 0.85 percent saline and commenced an observational case study on patients with frequent ear infections and asthma attacks. In all instances, his patients experienced marked improvement of their symptoms.

In 1998, at the age of sixty-three, Dr. Jones felt it was too late in life for him to embark on the massive project of starting a new company to sell his nasal spray, so he suggested that his son, Nathan, undertake the venture. Nate was, at the time, a commercial diver in the Gulf of Mexico and happened to be looking for a change. With great energy, he threw himself into founding and growing the company, which he named XLEAR. As he got into the business, he learned something firsthand that his father had already discovered—namely, that America's academic medical establishment and public health agencies apparently had no interest in investigating the product's potential benefit. Like Drs. Semmelweis and Snow in the mid-19th century, Dr. Jones and his son, Nate, were confounded that a perfectly logical and commonsensical proposition was being willfully ignored.

The FDA categorized the product as a "cosmetic" and not a "drug" and therefore refused to consider any tests that could demonstrate its efficacy in *preventing* upper respiratory tract infections. This maddening semantic categorization excluded the agency from even considering a simple form of personal hygiene that could eliminate or reduce the severity of a common illness that diminished the quality of life and general performance of millions of Americans. This categorization also meant that all advertising claims were regulated by the Federal Trade Commission (FTC). In 2002, Nate received a letter from the FTC requesting a phone conference. In a May 25, 2022, memo, Nate documented his interactions with the Commission as follows:

> We called the FTC and their agent was courteous. He told us that, from what they could see, we were making claims that could not be supported by science. After about 45 minutes we understood what the ground rules were and how they worked. We had studies showing how xylitol blocked bacterial adhesion, so we could say that xylitol blocks bacterial adhesion. What we could not say was the next step, the obvious step—that xylitol [or anything else] could help to prevent ear infections.
>
> We wanted to make some claims for how Xlear helps to prevent ear infections and so we funded research showing how effective it was. Ironically, the results were so good that the editor of the pediatric journal refused to publish the paper because he didn't believe that a simple sugar molecule sprayed up the nose could produce such results. After getting blocked and not being able to publish them, we gave up on doing more research and just pushed forward with the claims we could make.
>
> We moved forward with this, and until July of 2020, we never had another run in with regulatory agencies of any type. We'd long understood that xylitol blocks bacteria from adhering, but we never questioned what it would do to a virus. It wasn't until COVID-19 came along in March of 2020 that we even thought about this. Soon after SARS- CoV-2 arrived in the United States, several doctors began telling their patients and even advising in public forums that nasal rinses—with saline, povidone iodine, and with our products—could help to prevent infection or at least reduce the viral load in the nose, where the trouble begins. In early 2020 we sent our nasal spray to the virology lab at Utah State University to test if it is virucidal. Something in our product did indeed destroy SARS-CoV-2 in vitro. At first we thought it was the xylitol, but after 3 rounds of tests we discovered it was the Grapefruit Seed Extract preservative (at only 0.2% of the solution) that was destroying the virus.[219]

We sent our reports to Health and Human Services, but got no useful reply. We then sent them to the FDA, which refused to consider them on the grounds that a "cosmetic"—however promising it may be to help stop a raging pandemic—cannot even be reviewed by the agency. Categorizing our product as a cosmetic precluded the agency from authorizing us to perform a clinical trial in order to determine if our product can help to prevent or treat COVID-19.

A study performed at the University of Tennessee showed that xylitol blocked SARS-CoV-2 adhesion to vero monkey kidney tissue, the industry standard.[220] We shared this report with the media, and shortly thereafter the FTC sent us a warning letter—this time for the mere act of sharing academic medical studies. They claimed the study did not meet true scientific standards. And so, we contracted a lab to perform the same study on human upper airway cells in vitro and presented the favorable results to the FTC. The Commission replied that we needed to conduct a large randomized controlled trial, but the FDA refused to authorize and acknowledge such a trial due to its categorization of our product as a cosmetic.

Around the same time, an ICU pulmonologist in Florida named Dr. Gustavo Ferrer told us he was treating COVID-19 patients and seeing great results with the daily administration of Xlear nasal spray. This pulmonologist sent us a report documenting the positive results of his case study, but by then we'd come to the frustrating and painful conclusion that our federal health agencies simply refused to consider any evidence that nasal hygiene could help to prevent and treat COVID-19. This left us trying to disseminate whatever information we could. We found the entire interaction perplexing in the extreme. Was the FTC simply incapable of understanding concept of what we were doing? Several high-quality studies supported the concept of nasal hygiene. The practice has been around for millennia, is perfectly safe, and has been advised by licensed medical doctors for decades. And yet, suddenly FTC lawyers began insisting that we are not allowed to share this information.

The commonsense aspect of our proposition is so plain that it raises the question: Is the FTC deliberately trying to suppress the concept of nasal hygiene? Why else could explain why the Commission has issued warning letters and filed lawsuits against companies in the nasal hygiene space such as Xlear, Navage, NielMed, etc.?

For months we tried to persuade the FTC to elucidate its guidelines for what we could and could not say. The Commission has persistently replied

> that it cannot act as our marketing or regulatory advisor. But if they can't tell us what the guidelines are, how will any attorney be able to explain it to us? Recently the FTC told us we are not allowed to make any claim for our product regarding COVID-19. We also cannot state that xylitol blocks adhesion of SARS-CoV-2 to nasal cells, even though we have studies showing that it does. After months of this back and forth and wasting our money, we concluded that we have to face the Commission in court and let a judge or jury decide what we can or cannot say about our product.

Nasal hygiene—especially with diluted povidone iodine—also became the target of a mendacious propaganda campaign asserting that the practice is *dangerous*. An article in the popular online journal *Health* was illustrative. Under the headline "Why Doctors Warn Against Using Betadine to Prevent COVID-19," the report quoted Drs. Cassandra M. Pierre at the Boston Medical Center and Amesh A. Adalja, a senior scholar at the Johns Hopkins Center For Health Security—the same institution that hosted the pandemic planning simulation, Event 201, on October 19, 2019:

> Ingesting povidone-iodine can pose health risks. According to Dr. Adalja, povidone-iodine is commonly used for a gargle for sore throats, but accidentally ingesting it—whether you swallow it by mouth or put it up your nose and it drips down your throat—could cause gastrointestinal upset. If too much is swallowed, you'll need to seek medical attention or contact a Poison Control Center.
>
> "High doses of povidone-iodine could also cause kidney problems," added Dr. Adalja, along with potentially interfering with thyroid function. Furthermore, according to Dr. Pierre, "It may tarnish the color of a person's mucus membranes and skin, and even cause pulmonary irritation and shortness of breath. There is, of course, another danger too: It's that people are using remedies like povidone-iodine instead of the vaccine—which is proven to work."
>
> "It's odd because there's so much evidence for the vaccine, but people are turning to povidone-iodine," said Dr. Adalja. The best, safest evidence-based methods for preventing COVID-19 (and staving off severe disease and hospitalization) remain vaccination and masking. "People are looking for a quick and easy fix," said Dr. Pierre. "But the quick and easy preventative measure that's actually safe and proven is getting the vaccine and continuing to wear a mask."[221]

With these pronouncements, Drs. Adalja and Pierre closely adhered to the same familiar script of disparaging *all* early treatment modalities while at the same time exhorting everyone to get the vaccine. In their world, there was only one solution to the complex public health problem presented by SARS-CoV-2—i.e., *vaccinate.*

CHAPTER 45

"The best investment I've ever made."

Just before Christmas 2021, McCullough came over to my place to spend the day discussing the events set forth in this book. He arrived in characteristically good spirits, even though he'd just been hit with the most outrageous attack. That morning, Medscape—a popular medical news agency—published a report titled "Physicians of the Year: Best and Worst." An e-mail blast of the report went out to doctors all over the country. The e-mail's subject was, "Worst Physicians of 2021: Who disgraced the profession? See who made the list and how many of them you know?"

By itself the subject was foul, and under normal circumstances McCullough wouldn't have opened the e-mail. However, as it arrived on the heels of his Joe Rogan interview, which had sparked a flurry of negative communiqués about him on the Internet, it couldn't be ignored.

He went through the report's slide show of the "Worst of 2021." The first eight were a rogues' gallery of doctors convicted of performing fraudulent, unnecessary surgery, mass murder, making false diagnoses while under the influence of drugs, sexual harassment, and assault. Peter clicked on slide number 9 and saw a photo of himself under the headline "Baylor Gets Restraining Order Against COVID Vaccine Skeptic Doc."[222] The report explained that "Baylor was the first institution to cut ties with McCullough, who promoted the use of unproven therapies for COVID-19 and questioned the effectiveness of COVID-19 vaccines."

As McCullough saw this, his entire career flashed before him, starting with the essential fact that he'd always taken great pains to care for his patients. His clinical practice and medical license had always been beyond reproach. Then there were all his years of intense research into cardiorenal medicine, an area in which he was the most published author in history. His former editorship of two major journals and his inaugural editorship of the textbook *Cardiorenal Medicine*. Then all the time and effort he'd spent over the last twenty-two months in developing and advocating early treatment for COVID-19 patients. This endeavor had culminated in the "McCullough Protocol" that was the basis of treating millions of patients across the globe, sparing God knows how many from hospitalization and death.

Now he was placed in a lineup of convicted criminals, the first of which was wearing an orange correctional facility jumpsuit. He thought of himself wearing a suit and tie and testifying before the US Senate or on *Tucker Carlson Today* and the *Joe Rogan Experience* about ways to help sick and frightened Americans. How could Medscape do this? Who put them up to it? Who paid them? McCullough could only wonder.

He told me this bad new as we sat down in my living room for coffee. I studied him and listened closely to his tone. There wasn't a hint of bitterness or self-pity in it. I reflected on the times we'd spent together since my May 19 studio interview and realized I'd never once heard him complain about what he was up against or the losses he'd incurred for following the dictates of his conscience. On the contrary, he'd always reported his setbacks in a spirit of bemused amazement.

Since I'd gotten to know him, I was often reminded of the British author Rudyard Kipling, whose work I'd not read since I was a boy. A few lines from his poem "if—" came to mind:

> If you can bear to hear the truth you've spoken
> Twisted by knaves to make a trap for fools,
> Or watch the things you gave your life to, broken,
> And stoop and build 'em up with worn-out tools:

I interpreted McCullough's attitude as the cheerful stoicism of one who, in spite of his losses, could take deep satisfaction in being his own man with a clear conscience.

Often in our conversations we discussed the question *Why?* Why were such vast resources devoted to suppressing early treatment? McCullough remarked that doctors were still allowed to exercise their clinical judgment and prescribe drugs off-label for every other condition—just not for COVID-19. This was stupendously baffling, and he wondered if only a metaphysical explanation such as that offered by his pastor was adequate. The therapies he advocated were all perfectly safe, their profiles having been established by billions of doses taken. So again, what was the harm in letting people try them, especially given that nothing else was offered?

At the time Medscape published its libelous report, the sum total of evidence that COVID-19 is a treatable disease was overwhelming. There was no question that the right combination of therapeutic agents could stop the disease's progression. If early treatment protocols had been quickly adopted as the standard of care in the United States, approximately 70 percent of the nation's COVID-19 fatalities could have been prevented. Just before Christmas 2021, that came to about 610,000 preventable deaths.

In light of this disturbing reality, the actors who organized the propaganda campaign against early treatment and who deliberately hindered sick people from receiving medicines should be under investigation for mass negligent homicide. For these actors to claim they suppressed the medicines because their efficacy had not been 100 percent proven is no defense. It is the equivalent of saying you didn't throw a life ring to a drowning man in high seas because you doubted it would save him in such a violent storm.

When presented with an imperiled human being who will likely suffer grave harm or die if you do nothing, you have a duty to try to help the person with whatever is at hand. If your surfing buddy is bitten by a shark and suffering arterial bleeding, you will try to stop the bleeding with a surfboard leash. It wouldn't occur to you to do nothing and simply watch your friend bleed out because you doubt your makeshift tourniquet will work.

Viral disease progression is less dramatic and far more a matter of probability than the progression to death caused by severe trauma. Nevertheless, in the case of COVID-19, it quickly became apparent that the probability of severe disease in certain high-risk groups is far higher than the risk of adverse effects from taking hydroxychloroquine, azithromycin, ivermectin, high-dose aspirin, and prednisone. Anyone of ordinary prudence who isn't brainwashed by propaganda is capable of understanding this. So why was early treatment of COVID-19 suppressed?

The totality of circumstances indicates the Bio-Pharmaceutical Complex suppressed early treatment because it wasn't in their vaccine business plan. Precisely because early treatment works, it was perceived as complicating and undermining the rationale for total commitment to their vaccine solution. There is nothing clandestine about the total vaccine agenda. The leaders of the Bio-Pharmaceutical Complex, consisting of the Gates Foundation, the World Economic Forum, CEPI, and Big Pharma have made it crystal clear. The Gates Foundation is fanatically devoted to promoting mass vaccination. Over the years, Bill Gates has made innumerable statements about vaccines that suggest he has an obsessive interest in them. In a 2010 TED Talk about what will be required to reduce CO2 emissions, he addressed the factor of world population:

> First, we've got population. The world today has 6.8 billion people. That's headed up to about 9 billion. Now if we do a really great job on new vaccines, healthcare, reproductive health services, we could lower that by perhaps ten or fifteen percent.[223]

It is difficult if not impossible to understand how vaccines—which are supposed to prevent people from dying of infectious diseases—will help to reduce the world's population. It is also notable that Gates linked vaccines not only to public health, but also to his other great project for humanity and planet Earth—reducing carbon emissions.

In 2010, Gates proclaimed at the WEF's annual meeting in Davos, "We must make this the decade of vaccines."[224] To pursue this objective, his foundation collaborated with the WHO, UNICEF, and NIAID "to increase coordination across the international vaccine community and create a Global Vaccine Action Plan."[225] Nine years later—at the January 2019 WEF meeting in Davos—he announced his foundation's return on the $10 billion it had invested in vaccine development.

"We feel there's been over a 20-to-1 return, yielding $200 billion over those twenty or so years," he told CNBC's Betty Quick on *Squawk Box*.[226] This was, as he'd written in an essay for the *Wall Street Journal* the week before, "The Best Investment I've Ever Made."[227]

In January 2019, Gates was the patron saint of the Bio-Pharmaceutical Complex. His public image had come a long way since 1998, when the United States sued Microsoft for violating the Sherman Antitrust Act of 1898. The government successfully argued that Microsoft had abused its monopoly power with its operating system and web browser integration at the grave expense of several competitors. Gates's videotaped deposition on August 27, 1998, was leaked to the public, and many remarked that it was a masterpiece of sullen arrogance and petulance.[228] The video recordings "show a mogul who is incredulous that the government would dare to obstruct his route to world domination," as Professor John Naughton wrote in an opinion for *The Guardian*.[229] Gates was, it seemed, a man unaccustomed to being challenged by anyone.

Presiding US District of Columbia Judge Thomas Penfield Jackson wrote of the defendant as follows:

> Microsoft is a company with an institutional disdain for both the truth and for rules of law that lesser entities must respect. It is also a company whose senior management is not averse to offering specious testimony to support spurious defenses to claims of its wrongdoing.[230]

At the time Judge Jackson wrote this opinion, it wasn't controversial. In 1998, the 43-year-old Gates—the world's richest man—was frequently compared to John D. Rockefeller, widely regarded as the most ruthless monopolist of all time. Reflecting on Judge Jackson's opinion in a June 2000 essay, University of Illinois history Professor Richard Jensen referred to Gates as "the new Rockefeller" for his strong-arming of PC companies "in order to squelch the competitive threat posed by Netscape."[231]

At about the same time Professor Jensen penned his essay, Gates founded the Bill & Melinda Gates Foundation. He modeled it after the Rockefeller Foundation, which was started by John D. Rockefeller Sr. and his son John D. Rockefeller Jr., along with their advisor, Frederick Taylor Gates (no relation to Bill) in 1913. In recent years the Gates and Rockefeller Foundations have collaborated on many international public health issues. In 2016, a major study of their activities was published by the independent Global Policy Forum. As the *Guardian* reported the study:

> Using their immense wealth and influence with political and scientific elites, organisations like the Bill and Melinda Gates Foundation, the Rockefeller Foundation and others are promoting solutions to global problems that may undermine the UN and other international organisations, says the report . . .
>
> "Through the sheer size of their grant-making, personal networking and active advocacy, large global foundations have played an increasingly active role in shaping the agenda-setting and funding priorities of international organisations and governments. . . . Through their multiple channels of influence, the Rockefeller and Gates foundations have been very successful in promoting their market-based and bio-medical approaches towards global health challenges in the research and health policy community—and beyond."[232]

In the same year that Bill Gates started his foundation, he invited NIAID Director Anthony Fauci to visit him at his $127 million mansion on Lake Washington. As Fauci later recalled the occasion:

> Melinda was showing everyone on a tour of the house. And he said, "Can I have some time with you in my library," this amazingly beautiful library. . . . And it was there that he said, "Tony, you run the biggest infectious disease institute of the world. And I want to be sure the money I spend is well spent. Why don't we really get to know each other? Why don't we be partners?"[233]

The partnership between Bill Gates and Anthony Fauci—along with their ally the Wellcome Trust—controls somewhere around 57 percent of global biomedical research funding.[234]

The Gates Foundation also wields vast influence in the mass media. On August 21, 2020, the *Columbia Journalism Review* published a long, meticulous report titled "Journalism's Gates Keepers." The reporter, Tim Schwab, found $250 million of Gates Foundation grants to radio and television newsrooms (including $17.5 million to NPR alone), newspapers, journalism organizations, and advertising firms that create content. The Foundation has also given grants to the Poynter Institute and Gannett, whose *PolitFact* and *USA Today* fact-checking departments have defended Gates from what they characterize as "false conspiracy theories" and "misinformation" about his influence over public health policy. As the author pointed out, "The full scope of Gates's giving to the news media remains unknown because the foundation only publicly discloses money awarded through charitable grants, not through contracts."[235]

The Gates-Fauci partnership also works closely with major pharmaceutical companies, which have well-documented histories of suppressing old but effective generic drugs in order to promote new, commercially valuable medications. A 2017 article in the *Harvard Business Review* titled "How Pharma Companies Game the System to Keep Drugs Expensive" enumerated the tricks of the trade.[236] This report derived much of its data from a 2016 study published in the *Blood* medical journal. Reviewing the industry's influence in shaping public perceptions and policy, the authors wrote:

> From 1998 to 2013, pharmaceutical lobbying interests were 42% larger than the second highest-paying industry (health insurance). The $2.7 billion effort . . . almost equaled the combined contributions of Big Oil ($1.3 billion) and the defense industry ($1.5 billion). An even greater financial commitment is made

> to advertising. The United States and New Zealand are the only 2 countries that allow prescription medications to be advertised on television. In 2012, nearly $3.5 billion was invested in the United States in pharmaceutical marketing. For every dollar spent on research, an average of >$2 (sometimes up to $19) is spent on marketing. Nine out of 10 large pharmaceutical companies spend more on marketing than on research and development.[237]

For many years, Dr. Peter Breggin was a regular guest on popular TV shows, including the *Oprah Winfrey Show*, to talk about his work as a reformer. The invitations abruptly ceased in 1997, when the FDA issued permissive guidelines for "direct-to-consumer" advertising of pharmaceuticals on TV.

Pharmaceutical companies have also become extremely adept at "capturing" the men and women who are tasked with regulating them.[238] Just as FDA supervisor Dr. Curtis Wright got a job at Purdue after he approved OxyContin, FDA Commissioner Dr. Scott Gottlieb was appointed to Pfizer's Board of Directors shortly after he left the agency. Likewise, six months after he gave approval to Moderna's new COVID-19 vaccine, FDA Commissioner Dr. Stephen Hahn was offered a position at the venture capital firm that was one of Moderna's primary backers. Among major media players, Jim Smith, the CEO of Reuters, was appointed to Pfizer's Board of Directors in 2014 and is also on the international business council of the World Economic Forum.

Moncef Slaoui was the head of vaccine development at GlaxoSmithKline until 2017. At the time President Trump appointed him to lead Operation Warp Speed, he was a board member of both CEPI and Moderna, a primary candidate for Warp Speed funds. Though he resigned from the Moderna board to avoid a conflict of interest, he retained his stock options, which gained $2.4 million in value on the day the company announced favorable preliminary results of its Phase I trials. This raised concerns about his neutrality in judging its vaccine's safety and efficacy data, so he agreed to divest his shares of Moderna stock.[239]

In the final analysis, the Gates-Fauci-CEPI-Wellcome Trust-Big Pharma network controls or strongly influences:

1. Fauci's bully pulpit as the nation's most high-profile health advisor.
2. NIAID and other federal agencies.
3. University researchers who receive their grant funding.
4. Politicians who receive their campaign contributions.
5. Television networks that receive their advertising money.

6. Mass media outlets that receive Gates Foundation grants.

All of the above is thoroughly documented. Researchers all over the world, examining innumerable primary source documents, many published by the organizations themselves, have independently drawn the same conclusions. The facts of these networks and their immense financial power and influence are out in the open and plain to see for anyone who wishes to look.

In keeping with CEPI's business plan, when SARS-CoV-2 arrived, Bill Gates, Anthony Fauci, and their network proclaimed there were *no* treatments and that everyone needed to stay home until new vaccines were developed and every human on Earth received them. Because these vaccines were presented as the panacea to a global emergency, those who developed them received the following advantages:

1. Massive emergency R&D funding from government treasuries.
2. Massive purchasing commitments from the same governments.
3. Zero product liability.
4. Zero marketing costs.
5. An absolute global market (every man, woman, and child on Earth).
6. A recurring revenue stream from regular booster shots.

The COVID-19 Pandemic was "The Opportunity" to springboard mRNA vaccine technology from R&D labs, where it had been on the shelves for years, onto the world stage in as little as nine months. Without such an emergency and the all-powerful mechanism of "emergency countermeasures," the process would have probably taken another decade or more. CEPI's "innovative global partnership" did everything in its power to take advantage of this opportunity.

A key objective of their endeavor was the elimination of anything that could diminish the perceived urgent need for new vaccines. Hydroxychloroquine and ivermectin were already FDA-approved for other conditions and had well-established safety profiles, so they had to be relentlessly maligned. Even the arrival of safe and effective monoclonal antibodies—made by Lilly, Regeneron, and Glaxo-SmithKline—were suppressed by consigning them to obscurity. No major media reports. No billboard displays or local news channel coverage. Infusion centers were hidden from view as an awakening public sought these products that had equal EUA status with the vaccines.

Instead, every channel, billboard, and automated pharmacy phone line was jammed with COVID-19 vaccine announcements. At every opportunity and with every detected viral mutation, the government withdrew monoclonal antibodies from the market, even though no clinical failures were observed. President Biden proclaimed, "Get vaccinated!"—a useless dog command for senior citizens sweating with fever, gasping for breath, and desperately seeking monoclonal antibodies and other medicines that had helped former President Trump to recover from COVID-19.

The propaganda campaign against early treatment was highly effective on much of the population, which resulted in hundreds of thousands of preventable deaths and untold suffering. Suppressing treatment was the equivalent of disarming the people. Without these weapons to fight the virus, the citizenry was greatly reduced in its ability to go about its business and retain its freedom of choice. This action was consistent with what Dr. Peter Breggin concluded was the primary motive for suppressing early treatment and mandating the new vaccines—namely, seizing power over the entire human race.

In Dr. Breggin's view, the leaders of the Bio-Pharmaceutical Complex are "Global Predators" who are chiefly interested in power. Constantly augmenting their fortunes gives them a sense of personal power and enables them to obtain real power in the world. Money makes the world go round. A person who earns the US median household income of $70,000 per year would have to work for 1.85 million years, without spending a dime or losing one to inflation, in order to reach Bill Gates's fortune of $130 billion.

Throughout history, ambitious men from Alexander the Great to the Roman Emperor Trajan to Genghis Khan to Napoleon have sought world domination. Our era is no different. Now Bill Gates, Klaus Schwab, and their club of billionaire friends want to rule the world. Instead of commanding through military prowess, they rule through their control of information, technology, money, and powerful networks assembled over the last thirty years. Characterizing this proposition as a "conspiracy theory" takes us back to Max Frisch's play, *Biedermann and the Arsonists.* Not only is it obvious that WEF members aspire to rule the world; the organization *announced* its ambition at its 50th annual meeting in June 2020, which it called the Great Reset. Wikipedia describes the agenda as follows:

> The World Economic Forum generally suggests that a globalized world is best managed by a self-selected coalition of multinational corporations, governments and civil society organizations (CSOs). It sees periods of global

> instability—such as the financial crisis and the COVID-19 pandemic—as windows of opportunity to intensify its programmatic efforts.

As the asset-owning class learned during the Financial Crisis of 2008–09, a global crisis will trigger the world's central banks to generate trillions of dollars and distribute these funds to well-positioned companies and organizations in order to deal with the crisis. All of the WEF's "Strategic Partners" were well positioned to benefit from "emergency countermeasure" funding when the pandemic arrived, and most of the people who run these companies greatly increased their personal wealth during the pandemic.

The WEF's website features a page dedicated to the Great Reset on which is written:

> "The pandemic represents a rare but narrow window of opportunity to reflect, reimagine, and reset our world."—Professor Klaus Schwab, Founder and Executive Chairman, World Economic Forum.[240]

Bill Gates, Klaus Schwab, and their WEF cronies perceived long ago that an infectious disease pandemic provided the perfect mechanism for achieving their Great Reset ambition. By co-opting Bio-Pharmaceutical scientists, they positioned themselves to claim that only they possessed the knowledge and technology needed to solve the emergency confronting all mankind. Many scientists, eager for money and status, were all too willing to participate in their scheme. Their actions reminded Dr. Breggin of the many times he'd seen medical and pharmaceutical men try to achieve their grandiose ambitions by claiming to possess indisputable scientific authority.

With most of humanity stricken with mortal fear and unable to think rationally, these "Experts" presented themselves as the saviors of mankind and effectively hijacked public policy. No politician—not even Donald Trump—had the confidence to challenge their perceived authority. Eventually it became apparent to increasing numbers of people that their narrative was full of falsehoods. However, because the ensuing battle was fought on the grounds of medical authority, only medical doctors like Peter McCullough were in a position to fight back.

CHAPTER 46:

Bringing America Home

The morning of January 23, 2022, was very cold in Washington, DC. As I got out of a taxi near the Washington Monument, I realized I'd not dressed warmly enough for the Defeat the Mandates march. A group was gathered around the base as I arrived at about 10:30 a.m. I looked up at the soaring, 500-foot obelisk, and it occurred to me that George Washington would not have considered it a fitting monument. He was always a man of modesty and understatement.

It had been a long time since I'd thought about Washington. Standing there at his memorial, I realized that a quarter of a century had passed since I'd written a graduate school paper on his Farewell Address of 1796. Wistfully I wondered where the years had gone. So much had changed since I'd completed my formal education in 1996. Back then, the recent American Cold War victory had seemed like a triumphant vindication of our political system in which state power is precisely defined and limited, and individual liberty is the primary value. The US Constitution had seemed unassailable. Censorship and segregation of citizens for refusing an experimental medical procedure had been unthinkable. If someone had told me back then that twenty-five years hence, I would go to Washington to protest such a thing, I would have thought him insane.

Much of the press coverage of the announced event suggested that protesting COVID-19 vaccine mandates was a kooky, "fringe," "far right" thing to do. How had our government, mainstream media, and much of our people become so confused? Most of the other marchers with whom I spoke were not "antivaccine," and more than a few had received COVID shots. Like me, they opposed the state forcing people to receive injections of an experimental substance that had been developed at "warp speed" by companies (one with a felony conviction) with billions to gain and nothing to lose (no product liability).

Two months earlier, Bayer Pharmaceuticals Division President Stefan Oelrich gave a speech to the World Health Summit in which he remarked:

> I always like to say: If we had surveyed two years ago the public, "Would you be willing to take a gene or cell therapy and inject it into your body?" we probably would have had a 95% refusal rate. I think this pandemic has opened many people's eye to this innovation.[241]

This was a revealing statement. He clearly meant it as an endorsement for the massive psychological operation, successfully waged by the Bio-Pharmaceutical Complex, to condition humanity to accept innovative gene or cell therapy injections. He correctly deduced that "two years ago" most people would be reluctant to be injected with such products, especially if they understood what was in them. The "innovation" of which he spoke is a complete departure from conventional vaccines that had been around for almost a century. The new mRNA gene therapy shot does *not* use a dead or attenuated virus, but a segment of the virus's genome that programs human cells to produce uncontrolled amounts of the virus's pathogenic spike protein. If 95 percent of humanity understood this, and was still in possession of their common sense, they would have likely rejected the new gene therapy shots until far more safety data were available. Two years of relentless fear and propaganda had conditioned somewhere around 70 percent of humanity to embrace the new gene therapy shots.

Now we—the approximately 20,000 people who gathered for the march—were depicted as the wacky ones for protesting not the genetic injections themselves, but their forcible imposition on us by state power. Many of the unvaccinated had already lost their jobs, and none of us were allowed to enter restaurants or stay in hotels in our nation's capital. Adding to the absurdity, the experimental gene therapies did *not* prevent infection and transmission of the virus. This revelation apparently created a great cognitive dissonance among many who'd gotten the injections with the understanding they conferred true immunity. Now they were told that they weren't immunized against the virus, but merely protected from severe illness. This assurance was not entirely comforting, which may have partly explained why many vaccinated remained frightened of the unvaccinated.

None of the above conformed with long-established principles of immunology or logic.

I Googled Washington's "Farewell Address" on my phone and quickly found his passage on faction:

> The alternate domination of one faction over another, sharpened by the spirit of revenge . . . has perpetrated the most horrid enormities, is itself a frightful despotism.[242]

We, the marchers, were the minority faction, and we were being treated as mentally deficient, public health hazards, unworthy of possessing the full rights of citizens. Much of the press had insinuated that we were an unruly bunch, given to violent disorder. Marching down the Mall toward the Lincoln Memorial, I marveled at our restraint, given the frightful despotism that was being imposed upon us.

We arrived at the Memorial with a full program of speeches ahead of us. I saw McCullough, Robert Malone, Pierre Kory, and others, standing together on the stage. The sight of them reminded me of Henry V's rousing words—"We few, we happy few, we band of brothers"—and it lifted my spirits. The trouble was, the enemy wasn't a foreign army arrayed before us on a battlefield, but the irrationality that had seized our own people. With an identifiable physical danger, at least you know what you're up against. Not so with hysteria, which spreads like a contagion, rendering humans capable of doing just about anything, no matter how terrible. As the German poet Friedrich Schiller had put it:

> It's dangerous to wake the lion,
> Dreadful is the tiger's tooth.
> But the most terrible of terrors,
> Is man in his madness.[243]

This wasn't the time and place to entertain such dreary thoughts. Maybe I was just tired and cold. I saw some food trucks parked on the street flanking the left side of the Memorial, so I wandered over and found one serving coffee. Just as I was paying for it, I heard McCullough's amplified voice, starting his speech, so I ran back to hear it.

Almost two years had elapsed since he'd embarked on his quest to discover and advocate the early treatment of COVID-19. He'd started his journey as a doctor in Dallas, just trying to fulfill his Hippocratic Oath. Now he was the central figure in a global conflict, standing on the most iconic public speaking stage on Earth. When in history, I wondered, had medical authority been the decisive credential of a man advocating for liberty—the cornerstone of the American way of life, and of all nations that call themselves liberal democracies?

The colossal statue of Abraham Lincoln, looming behind him, reminded me of the sublime brevity of the Gettysburg Address, and I hoped that McCullough's remarks would be similarly structured and succinct. He didn't disappoint:

> The efforts put forward by the best doctors in America to provide compassionate care for their patients were impeded, and it amplified fear, suffering, hospitalization, and death. And it affected the most vulnerable people in our society—our seniors, our people of color in this country. They were adversely affected in the most severe and in some cases the most final way. . . . By treating the disease, we reduce the intensity and duration of symptoms, and by that mechanism we reduce the probability of hospitalization and death. We have heard that we should stay in lockdown, wear a mask, social distance, and wait for a vaccine. Now there's not a single person here [on stage] who is against the broad use of vaccines as we use them in our clinical practice, myself included. But when the [COVID-19] vaccines were in development, we knew it was a gamble of a lifetime if not a gamble of all time for humanity. . . .
>
> We have three critical circles that all of you are in today. We have a circle of medical freedom. You have the freedom. You and you alone have the autonomy over your body. You have the freedom to determine what happens to your body. This is solely your possession. *In many ways it's the only thing you really have.* That circle of medical freedom is inextricably linked to the circle of social freedom, and that social freedom includes your family, your employment, your faith, and your citizenry in any country. And that is inextricably linked to your circle of economic freedom. Medical freedom is linked to social and economic freedom. If we allow the circle of medical freedom to be touched, let alone broken, all of the circles fracture. . . .
>
> The writing is on the wall, and the determination to preserve medical freedom is in your hands. It can't be any clearer. This moment cannot be more decisive. So, join me, and join these heroic doctors and nurses and others in helping bring America home. Thank you.

Notes

Prologue

1. Johns Hopkins Center for Health Security, EVENT 201, October 18, 2019. https://www.centerforhealthsecurity.org/event201/
2. Ibid. https://www.centerforhealthsecurity.org/event201/videos.html

Chapter 1: The Plague Is Coming

3. H. J. Lane et al. Oliver Wendell Homes (1809–1894) and Ignaz Philipp Semmelweis (1818–1865): Preventing the Transmission of Puerperal Fever. *American Journal of Public Health,* Sept. 20, 2011. https://ajph.aphapublications.org/doi/10.2105/AJPH.2009.185363
4. Jen Pinkowski. The 1918 Pandemic Mistake that Changed Medicine Forever. *National Geographic*, March 21, 2022. https://www.nationalgeographic.com/science/article/the-1918-pandemic-mistake-that-changed-medicine-forever
5. HV Epps. Influenza: Exposing the true killer. *J Exp Med.* 2006 Apr 17; 203(4): 803. https://www.ncbi.nlm.nih.gov/pmc/articles/PMC2118275/
6. N. Petrosillo et al. COVID-19, SARS and MERS: are they closely related? *Clinical Microbiology and Infection*. March 28, 2020. https://www.clinicalmicrobiologyandinfection.com/article/S1198-743X(20)30171–3/fulltext
7. M. J. Vincent et al. Chloroquine is a potent inhibitor of SARS coronavirus infection and spread. *Virology Journal,* volume 2, Article number: 69 (2005) https://virologyj.biomedcentral.com/articles/10.1186/1743-422X-2-69?fbclid=IwAR2apSvsdRRkpcGGA3Pk0smL4AJeEgu9UI_xfB0ghT4QVTwHo8L2P6EPkak
8. CDC. Partner Key Messages on the 1918 Influenza Pandemic Commemoration https://www.cdc.gov/flu/pandemic-resources/1918-commemoration/key-messages.htm
9. G Onder et al. Case-Fatality Rate and Characteristics of Patients Dying in Relation to COVID-19 in Italy. *JAMA*. 2020;323(18):1775–1776. doi:10.1001/jama.2020.4683, March 23, 2020. https://jamanetwork.com/journals/jama/fullarticle/2763667
10. Laura J. Nelson et al. Los Angeles Beaches reopen in a coronavirus milestone, *Los Angeles Times.* May 13, 2020. https://news.yahoo.com/los-angeles-county-beaches-reopen-144237831.html
11. Woody Allen. *Bananas*, 1971.

Chapter 2: Preparing for Battle

12. Nurse Who Caught Ebola Settles Suit Against Dallas Hospital. *NBC News/Source Associated Press*, Oct. 24, 2016. https://www.nbcnews.com/storyline/ebola-virus-outbreak/nurse-who-caught-ebola-settles-suit-against-dallas-hospital-n672081

Chapter 3: Gandalf of Marseille

13. Philippe Chapelin. The legacy of French physician and microbiologist Didier Raoult: genealogical investigation. *GeFrance*, May 29, 2020. https://gefrance.com/the-legacy-of-french-physician-and-microbiologist-didier-raoult-genealogical-investigation/
14. Scott Sayare. He was a Science Star. Then He Promoted a Questionable Cure for COVID-19. *New York Times*, May 12, 2020. https://www.nytimes.com/2020/05/12/magazine/didier-raoult-hydroxychloroquine.html
15. Karl Laske and Jacque Massey. The strange saga of how France helped build Wuhan's top-security virus lab. *MEDIAPART*, May 31, 2020. https://www.mediapart.fr/en/journal/france/310520/strange-saga-how-france-helped-build-wuhans-top-security-virus-lab?_locale=en&onglet=full
16. 'The Never Again Plan': Moderna CEO Stéphane Bancel wants to stop the next COVID-19—before it happens. *Advisory Board*, Dec. 22, 2020. https://www.advisory.com/blog/2020/12/moderna-ceo-COVID-vaccine-bancel
17. Etienne Campion. Didier Raoult et le milieu médical parisien : histoire d'une détestation réciproque. *Marianne*, March 26, 2020. https://www.marianne.net/politique/didier-raoult-et-le-milieu-medical-parisien-histoire-d-une-detestation-reciproque
18. Ibid.
19. J. Gao et al. Breakthrough: Chloroquine phosphate has shown apparent efficacy in treatment of COVID-19 associated pneumonia in clinical studies. *BioScience Trends*. 2020; 14(1):72–73, February 19, 2020. https://www.jstage.jst.go.jp/article/bst/14/1/14_2020.01047/_pdf/-char/en
20. P. Gautret et al. Hydroxychloroquine and azithromycin as a treatment of COVID-19: results of an open-label non-randomized clinical trial. *Int J Antimicrob Agents*. 2020 Jul; 56(1): 105949. https://www.ncbi.nlm.nih.gov/pmc/articles/PMC7102549/
21. Norman Doidge. Hydroxychloroquine: A Morality Tale. *Tablet Magazine*. Aug. 13, 2020. https://www.tabletmag.com/sections/science/articles/hydroxychloroquine-morality-tale
22. Kwak Sun-Sun. Physicians Work Out Treatment Guidelines for Coronavirus. *Korea Biomedical Review*, February 13, 2020. http://www.koreabiomed.com/news/articleView.html?idxno=7428
23. J. M. Todaro, MD and G. J. Rigano, Esq. An Effective Treatment for Coronavirus (COVID-19). *Google Docs*, March 13, 2020. https://docs.google.com/document/d/e/2PACX-1vTi-g18ftNZUMRAj2SwRPodtscFio7bJ7GdNgbJAGbdfF67WuRJB3ZsidgpidB2eocFHAVjIL-7deJ7/pub
24. Elon Musk. Maybe worth considering chloroquine for C19. *Twitter*, March 16, 2020. https://twitter.com/elonmusk/status/1239650597906898947

Chapter 4: A Vaccine in Record Speed

25. NIAID. NIH Clinical Trial of Investigational Vaccine for COVID-19 Begins. March 16, 2020. https://www.niaid.nih.gov/news-events/nih-clinical-trial-investigational-vaccine-COVID-19-begins
26. CDC. First Travel-related Case of 2019 Novel Coronavirus Detected in United States. January 21, 2020. https://www.cdc.gov/media/releases/2020/p0121-novel-coronavirus-travel-case.html
27. CDC. CDC Museum COVID-19 Timeline. https://www.cdc.gov/museum/timeline/COVID19.html
28. Moderna. Our Story. https://www.modernatx.com/about-us/our-story?
29. Alexander Tin. Moderna offers NIH co-ownership of COVID vaccine patent amid dispute with government. *CBS News*, Nov. 15, 2021. https://www.cbsnews.com/news/moderna-COVID-vaccine-patent-dispute-national-institutes-health/

Chapter 5: "The Opportunity"

30. CEPI. ARTICLES OF ASSOCIATION. https://cepi.net/wp-content/uploads/2019/07/CEPI-Articles-of-Association_web.pdf
31. CEPI. OUR MISSION. https://cepi.net/about/whyweexist/
32. CEPI. Preliminary Business Plan 2017–2021. https://cepi.net/wp-content/uploads/2019/02/CEPI-Preliminary-Business-Plan-061216_0.pdf

Chapter 6: Unwishful Thinking

33. Merlin Wilcox et al. *Traditional Medicinal Plants and Malaria*. Chapter 2: Cinchona, New York: CRC Press, 2005. https://books.google.com/books?id=L3lZiwsCZoYC&pg=PA23#v=onepage&q&f=false
34. Norman Doidge. Hydroxychloroquine: A Morality Tale. Tablet Magazine. August 13, 2020. https://www.tabletmag.com/sections/science/articles/hydroxychloroquine-morality-tale
35. I. Ben-Zvi et al. Hydroxychloroquine: From Malaria to Autoimmunity. *Clin Rev Allergy Immunol.* 2012; 42(2): 145–153 https://www.ncbi.nlm.nih.gov/pmc/articles/PMC7091063/
36. A. J. W. te Velthuis et al. Zn2+ Inhibits Coronavirus and Arterivirus RNA Polymerase Activity *In Vitro* and Zinc Ionophores Block the Replication of These Viruses in Cell Culture. *PLOS PATHOGENS*, November 4, 2010. https://journals.plos.org/plospathogens/article?id=10.1371/journal.ppat.1001176
37. Amber Dance. The shifting sands of "gain-of-function" research. *Nature*, October 27, 2021. https://www.nature.com/articles/d41586-021-02903-x
38. UNC Gillings School of Global Public Health. Carolina research produces effective experimental pill to treat COVID-19. October 1, 2021. https://sph.unc.edu/sph-news/carolina-research-produces-effective-experimental-pill-to-treat-COVID-19/
39. J. Dyall et al. Repurposing of Clinically Developed Drugs for Treatment of Middle East Respiratory Syndrome Coronavirus Infection. *Antimicrob Agents Chemother*. 2014 Aug; 58(8): 4885–4893. https://www.ncbi.nlm.nih.gov/pmc/articles/PMC4136000/

40. J. Liu et al. Hydroxychloroquine, a less toxic derivative of chloroquine, is effective in inhibiting SARS-CoV-2 infection in vitro. *Nature*, March 18, 2020. https://www.nature.com/articles/s41421-020-0156-0
41. Donald Trump Coronavirus Task Force Briefing Transcript. March 19, 2020 https://www.rev.com/blog/transcripts/donald-trump-coronavirus-task-force-briefing-transcript-march-19-trump-takes-shots-at-the-media
42. Donald Trump Coronavirus Task Force Briefing Transcript. March 20, 2020 https://www.rev.com/blog/transcripts/donald-trump-coronavirus-task-force-march-20-press-conference-transcript-trump-spars-with-reporters-in-fiery-briefing
43. Editorial Board Opinion: Trump is spreading false hope for a virus cure—and that's not the only damage. *Washington Post*, March 25, 2020. https://www.washingtonpost.com/opinions/global-opinions/trump-is-spreading-false-hope-for-a-virus-cure--and-thats-not-the-only-damage/2020/03/25/587b26d8-6ec3-11ea-b148-e4ce3fbd85b5_story.html
44. Bill Gates. Opinion: Here's how to make up for lost time on COVID-19. *Washington Post*, March 31, 2020. https://www.washingtonpost.com/opinions/bill-gates-heres-how-to-make-up-for-lost-time-on-COVID-19/2020/03/31/ab5c3cf2-738c-11ea-85cb-8670579b863d_story.html
45. Siemny Kim. Bill Gates says foundation will invest billions in fight to stop COVID-19. *KIRO 7 News*, April 6, 2020. https://www.kiro7.com/news/local/bill-gates-says-foundation-will-invest-billions-fight-stop-COVID-19/MMAFTSVGKZHPTEGYKEQKMRWTWU/
46. Bill Gates. What you need to know about the COVID-19 vaccine. *GatesNotes*, April 30, 2020. https://www.gatesnotes.com/health/what-you-need-to-know-about-the-COVID-19-vaccine

Chapter 7: The "Simple Country Doctor"

47. Coronavirus Epidemic Update 34. *MedCram*, March 10, 2020. https://www.youtube.com/watch?v=U7F1cnWup9M&t=171s

Chapter 8: "My detractors are children!"

48. M. Million et. al. Early treatment of COVID-19 patients with hydroxychloroquine and azithromycin: A retrospective analysis of 1061 cases in Marseille, France. *Travel Med Infect Dis.* 2020 May-June; 35: 101738. May 5, 2020. https://www.ncbi.nlm.nih.gov/pmc/articles/PMC7199729/
49. Catherine Paley-Vincent and Boudet-Gizardin. Can hydroxychloroquine be legally prescribed in France for COVID-19? *Ginestie Magellan Paley-Vincent Law Firm.* https://www.ginestie.com/en/COVID-19-can-hydroxychloroquine-be-legally-prescribed-in-France-for-patients-with-COVID-19/
50. J. Magagnoli et al. Outcomes of hydroxychloroquine usage in United States veterans hospitalized with COVID-19. *medRxiv preprint.* April 21, 2020. https://www.medrxiv.org/content/10.1101/2020.04.16.20065920v1.full.pdf
51. Elizabeth Cohen and Migali Nigam, MD. Study finds no benefit, higher death rate in patients taking hydroxychloroquine for COVID-19. *CNN*, April 21, 2020. https://www.cnn.com/2020/04/21/health/hydroxychloroquine-veterans-study/index.html

52. L. Tan et al. Lymphopenia predicts disease severity of COVID-19: a descriptive and predictive study. *Nature*, March 27, 2020. https://www.nature.com/articles/s41392-020-0148-4
53. Emilie Blachere, Entretien exclusif avec Didier Raoult: "Je suis un renégat." *Paris Match*. May 9, 2020. https://www.parismatch.com/Actu/Sante/Professeur-Didier-Raoult-Je-suis-un-renegat-1683722

Chapter 9: Memento Mori

54. Robert Mackey. Alarm and Confusion at Fox News as Trump Says He Takes Hydroxychloroquine. *The Intercept*, May 19, 2020. https://theintercept.com/2020/05/19/alarm-confusion-fox-news-trump-says-takes-hydroxychloroquine/
55. FDA Emergency Use Authorization for Vaccines Explained. Nov. 20, 2020. https://www.fda.gov/vaccines-blood-biologics/vaccines/emergency-use-authorization-vaccines-explained
56. Donald Trump Says He's Taking Hydroxychloroquine to Prevent Coronavirus in Press Conference. *Rev Transcripts*, May 18, 2020. https://www.rev.com/blog/transcripts/transcript-donald-trump-says-hes-taking-hydroxychloroquine-to-prevent-coronavirus-in-press-conference
57. Dava Sobel, *Longitude*. Bloomsbury: New York, 1995.
58. FDA, Understanding Unapproved Use of Approved Drugs "Off Label." Feb. 5, 2018. https://www.fda.gov/patients/learn-about-expanded-access-and-other-treatment-options/understanding-unapproved-use-approved-drugs-label

Chapter 10: Shooting the Message

59. F. Wolfe et al. Rates and predictors of hydroxychloroquine retinal toxicity in patients with rheumatoid arthritis and systemic lupus erythematosus. *Arthritis Care Res* (Hoboken) 2010 Jun;62(6):775–84. https://pubmed.ncbi.nlm.nih.gov/20535788/
60. Dr. Oz. Interview with Dr. Daniel Wallace. April 7, 2020. https://www.youtube.com/watch?v=htyCEeq_YVI
61. H. A. Risch, Early Outpatient Treatment of Symptomatic, High-Risk COVID-19 Patients That Should Be Ramped Up Immediately as Key to the Pandemic Crisis. *American Journal of Epidemiology*, Volume 189, Issue 11, November 2020, Pages 1218–1226. Published. 27 May 2020. https://academic.oup.com/aje/article/189/11/1218/5847586
62. Stephen J. Hatfill, MD. The intentional destruction of the National Pandemic Plan. *The Desert Review*, August 30, 2021. https://www.thedesertreview.com/opinion/columnists/the-intentional-destruction-of-the-national-pandemic-plan/article_7eb45834-09a8-11ec-a0ad-339a1fbfc2a6.html
63. The Rockefeller Foundation. Dr. Rick Bright. https://www.rockefellerfoundation.org/profile/dr-rick-bright/
64. C-SPAN, Universal Flu Vaccine. October 29, 2019. https://www.c-span.org/video/?465845–1/universal-flu-vaccine
65. BIT CHUTE. Dr. Zelenko Exposes How Dr. Rick Bright's Very Bad Move Sabotaged Early COVID Treatment Killing Masses. Sept. 2, 2021. https://www.bitchute.com/video/WHIktyjGEPcF

66. Stephen J. Hatfill, MD. The intentional destruction of the National Pandemic Plan. *The Desert Review*, August 30, 2021. https://www.thedesertreview.com/opinion/columnists/the-intentional-destruction-of-the-national-pandemic-plan/article_7eb45834-09a8-11ec-a0ad-339a1fbfc2a6.html
67. BIT CHUTE. Dr. Zelenko Exposes How Dr. Rick Bright's Very Bad Move Sabotaged Early COVID Treatment Killing Masses, Sept. 2, 2021. https://www.bitchute.com/video/WHIktyjGEPcF
68. Grace Segers. HHS ousts vaccine expert who pushed back on COVID-19 treatment. *CBS NEWS*, April 23, 2020. https://www.cbsnews.com/news/rick-bright-doctor-removed-hydroxychloroquine-health-human-services-coronavirus-COVID-19-treatment/
69. Podcast Episode #1757 Dr. Robert Malone, MD. *The Joe Rogan Experience*, Spotify, Dec. 2021. https://open.spotify.com/episode/3SCsueX2bZdbEzRtKOCEyT
70. Samantha McGrail. BARDA Gives $483M to Moderna for COVID-19 Vaccine Development. *PHARMANEWS INTELLIGENCE*, April 24, 2020. https://pharmanewsintel.com/news/barda-gives-483m-to-moderna-for-COVID-19-vaccine-development
71. Sheryl Gay Stolberg, Whistle-Blowing Scientist Quits Government With Final Broadside. *New York Times*, Oct. 6, 2020. https://www.nytimes.com/2020/10/06/us/politics/whistle-blower-rick-bright.html
72. Grace Segers. HHS ousts vaccine expert who pushed back on COVID-19 treatment. *CBS NEWS*, April 23, 2020. https://www.cbsnews.com/news/rick-bright-doctor-removed-hydroxychloroquine-health-human-services-coronavirus-COVID-19-treatment/

Chapter 11: "Cuomosexuals"

73. State of New York Executive Chamber. Executive Order. March 23, 2020. https://www.governor.ny.gov/sites/default/files/atoms/files/EO_202.10.pdf
74. New York State Department of Health. Advisory: Hospital Discharges and Admissions to Nursing Homes. March 25, 2020. https://skillednursingnews.com/wp-content/uploads/sites/4/2020/03/DOH_COVID19—NHAdmissionsReadmissions__032520_1585166684475_0.pdf
75. J. K. Louie et al. Rhinovirus Outbreak in a Long-Term Care Facility for Elderly Persons Associated with Unusually High Mortality. *Clinical Infectious Diseases*, Volume 41, Issue 2, 15 July 2005, Pages 262–265. https://academic.oup.com/cid/article/41/2/262/531713
76. G. Onder et al. Case-Fatality Rate and Characteristics of Patients Dying in Relation to COVID-19 in Italy. *JAMA*, March 23, 2020. https://jamanetwork.com/journals/jama/fullarticle/2763667
77. Jelena Dzhanova. New York state now has more coronavirus cases than any country outside the US. *CNBC*, April 10, 2020. https://www.cnbc.com/2020/04/10/new-york-state-now-has-more-coronavirus-cases-than-any-country-outside-the-us.html
78. Bernhard Condon et al. AP count: Over 4,500 virus patients sent to NY nursing homes. *AP News*, May 22, 2020. https://apnews.com/article/health-us-news-ap-top-news-weekend-reads-virus-outbreak-5ebc0ad45b73a899efa81f098330204c
79. Bernhard Condon and Jennifer Peltz. AP: Over 9,000 virus patients sent into NY nursing homes. *AP News*, Feb. 11, 2021. https://apnews.com/article/new-york-andrew-cuomo-us-news-coronavirus-pandemic-nursing-homes-512cae0abb55a55f375b3192f2cdd6b5

80. NY Attorney General. Attorney General James Releases Report on Nursing Homes' Response to COVID-19. January 28, 2021. https://ag.ny.gov/press-release/2021/attorney-general-james-releases-report-nursing-homes-response-COVID-19
81. Bernadette Hogan et al. Cuomo aide Melissa DeRosa admits they hid nursing home data so feds wouldn't find out. *New York Post*, Feb. 11, 2021. https://nypost.com/2021/02/11/cuomo-aide-admits-they-hid-nursing-home-data-from-feds/
82. Governor Cuomo embraces the term "Cuomosexual." *Los Angeles Times*, April 28, 2020. https://www.latimes.com/entertainment-arts/story/2020-04-28/andrew-cuomo-sexual-ellen-degeneres-youtube
83. Eric Lach. Andrew Cuomo's Downfall Began with a Book Deal. *New Yorker*, November 23, 2021. https://www.newyorker.com/news/our-local-correspondents/andrew-cuomos-downfall-began-with-a-book-deal
84. Ibid.
85. Colin Dwyer. Andrew Cuomo To Receive International Emmy For "Masterful" COVID-19 Briefings. *NPR*, November 21, 2020. https://www.npr.org/sections/coronavirus-live-updates/2020/11/21/937445923/andrew-cuomo-to-receive-international-emmy-for-masterful-COVID-19-briefings
86. Marisa Kwiatkowski. "A national disgrace": 40,600 deaths tied to US nursing homes. *USA TODAY*, June 1, 2020. https://www.usatoday.com/story/news/investigations/2020/06/01/coronavirus-nursing-home-deaths-top-40-600/5273075002/
87. Maggie Haberman and Jesse McKinley. How Cuomo's Team Tried to Tarnish One of His Accusers. *New York Times*, March 16, 2021. https://www.nytimes.com/2021/03/16/nyregion/cuomo-lindsey-boylan.html
88. State of New York, Office of the Attorney General, SUBPOENA AD TESTIFICANDUM THE PEOPLE OF THE STATE OF NEW YORK. https://ag.ny.gov/sites/default/files/chris_cuomo_exhibits_-_combined.pdf
89. NY Attorney General. Attorney General James Releases Report on Nursing Homes' Response to COVID-19. January 28, 2021. https://ag.ny.gov/press-release/2021/attorney-general-james-releases-report-nursing-homes-response-COVID-19
90. Lizzie Widdicombe. Diving into the Subconscious of the "Cuomosexual." *New Yorker*, August 7, 2021. https://www.newyorker.com/culture/annals-of-inquiry/diving-into-the-subconscious-of-the-cuomosexual
91. Christina Anderson. In the Coronavirus Fight in Scandinavia, Sweden Stands Apart. *New York Times*, March 28, 2020. https://www.nytimes.com/2020/03/28/world/europe/sweden-coronavirus.html

Chapter 12: The Wonder Drug

92. J. D. Druce et al. The FDA-approved drug ivermectin inhibits the replication of SARS-CoV-2 in vitro. *Antiviral Research* Volume 178 June 2020, 10478. https://www.sciencedirect.com/science/article/pii/S0166354220302011?via%3Dihub
93. A. Crump and S. Omura. Ivermectin, "Wonder drug" from Japan: the human use perspective. *Proc Jpn Acad Ser B Phys Biol Sci*. 2011 Feb 10; 87(2): 13–28. https://www.ncbi.nlm.nih.gov/pmc/articles/PMC3043740/
94. MERCK. Over 30 Years: The Mectizan Donation Program. January 6, 2021. https://www.merck.com/stories/mectizan/

95. The Nobel Prize. PRESS RELEASE. 2015-10-05. https://www.nobelprize.org/prizes/medicine/2015/press-release/
96. A. Crump. Ivermectin: enigmatic multifaceted "wonder" drug continues to surprise and exceed expectations. *Nature*, February 15, 2017. https://www.nature.com/articles/ja201711
97. C. Veeresham. Natural products derived from plants as a source of drugs. *J Adv Pharm Techlol Res* v. 3(4); Oct-Dec 2020. https://www.ncbi.nlm.nih.gov/pmc/articles/PMC3560124/
98. FDA. FDA Letter to Stakeholders: Do Not Use Ivermectin Intended for Animals as Treatment for COVID-19 in Humans. April 26, 2021. https://www.fda.gov/animal-veterinary/product-safety-information/fda-letter-stakeholders-do-not-use-ivermectin-intended-animals-treatment-COVID-19-humans
99. J. C. Rajter et al. Use of Ivermectin Is Associated With Lower Mortality in Hospitalized Patients With Coronavirus Disease 2019. *CHEST INFECTIONS: ORIGINAL RESEARCH*| VOLUME 159, ISSUE 1, P85-92, JANUARY 01, 2021. https://journal.chestnet.org/article/S0012-3692(20)34898–4/fulltext
100. P. Kory et. al. Review of the Emerging Evidence Demonstrating the Efficacy of Ivermectin in the Prophylaxis and Treatment of COVID-19. *American Journal of Therapeutics*: May/June 2021—Volume 28 - Issue 3—p e299-e318. https://journals.lww.com/americantherapeutics/fulltext/2021/06000/review_of_the_emerging_evidence_demonstrating_the.4.aspx
101. NIH. COVID-19 Treatment Guidelines. August 27, 2020. https://files.COVID19treatmentguidelines.nih.gov/guidelines/archive/COVID19treatmentguidelines-08-27-2020.pdf
102. FLCCC ALLIANCE. One Page Summary of the Clinical Trials Evidence for Ivermectin in COVID-19 as of January 11, 2021. https://COVID19criticalcare.com/wp-content/uploads/2020/12/One-Page-Summary-of-the-Clinical-Trials-Evidence-for-Ivermectin-in-COVID-19.pdf

Chapter 13: Dr. Fauci Goes to Bat for Remdesivir

103. Kathryn Ardizzone. Role of the Federal Government in the Development of Remdesivir. *KEI Briefing Note* 2020:1.(March 20, 2020). https://www.keionline.org/wp-content/uploads/KEI-Briefing-Note-2020_1GS-5734-Remdesivir.pdf
104. COVID-19 Treatment Guidelines Panel. Coronavirus Disease 2019 (COVID-19) Treatment Guidelines. National Institutes of Health. https://www.COVID19treatmentguidelines.nih.gov/about-the-guidelines/panel-financial-disclosure/
105. Robert F. Kennedy Jr. *The Real Anthony Fauci: Bill Gates, Big Pharma, and the Global War on Democracy and Public Health* (Children's Health Defense) (pp. 163–164). Skyhorse. Kindle Edition.
106. Ibid. (pp. 164–165)
107. Ibid. (p. 166)
108. Y. Wang, MD, et al., Remdesivir in adults with severe COVID-19: a randomised, double-blind, placebo-controlled, multicentre trial. the *Lancet*, Vol 395, I 10236, P1569-1578, (May 16, 2020), https://doi.org/10.1016/S0140-6736(20)31022–9

109. Phil Taylor. Gilead Slides as second Chinese trial of remdesivir is stopped. *Pharmaphorum*, April 15, 2020. https://pharmaphorum.com/news/gilead-slides-as-two-chinese-trials-of-remdesivir-are-stopped/
110. Laurel Wamsley and Carmel Wroth. Antiviral Drug Remdesivir Shows Promise For Treating Coronavirus In NIH Study. *NPR*, April 29, 2020. https://www.npr.org/sections/health-shots/2020/04/29/848034963/antiviral-drug-remdesivir-shows-promise-for-treating-coronavirus-in-nih-study
111. Iain Martin. U.S. Buys the World's Supply of Breakthrough Coronavirus Drug Remdesivir. *Forbes*, July 1, 2020. https://www.forbes.com/sites/iainmartin/2020/07/01/us-buys-the-world-supply-of-breakthrough-coronavirus-drug-remdesivir/?sh=1fe705755472
112. Francisco Guarascio. EU makes 1 billion-euro bet on Gilead's COVID drug before trial results. *REUTERS*, October 13, 2020. https://www.reuters.com/article/us-health-coronavirus-eu-remdesivir/eu-makes-1-billion-euro-bet-on-gileads-COVID-drug-before-trial-results-idUSKBN26Y25K
113. Virginie Joron. Question for written answer E-006511/2020 to the Commission, Rule 138. Subject: Remdesivir: the EUR 1 billion scandal of the fake treatment for COVID-19 purchased by the Commission. Parliamentary Questions, November 30, 2020. https://www.europarl.europa.eu/doceo/document/E-9-2020-006511_EN.html
114. Billions wasted over swine flu, says Paul Flynn MP. *BBC*, June 24, 2010. https://www.bbc.com/news/10396382
115. Katharine J. Wu and Gina Kolata. Remdesivir Fails to Prevent COVID-19 Deaths in Huge Trial. *New York Times*, October 15, 2020. https://www.nytimes.com/2020/10/15/health/coronavirus-remdesivir-who.html

Chapter 15: On the Front Line of Critical Care

116. CDC, "What is Sepsis?" https://www.cdc.gov/sepsis/what-is-sepsis.html
117. P. Marik et al. Hydrocortisone, Vitamin C, and Thiamine for the Treatment of Severe Sepsis and Septic Shock: A Retrospective Before-After Study *CHEST* 2017 Jun;151(6):1229–1238. doi: 10.1016/j.chest.2016.11.036. Epub 2016 Dec. https://pubmed.ncbi.nlm.nih.gov/27940189/
118. Michael Capuzzo. The Drug that Cracked COVID. *Mountain Home*, May 2021. https://www.mountainhomemag.com/article_tags/michael-capuzzo
119. Ibid.
120. US Senate Committee—Dr. Pierre Kory CRITICAL Witness Testimony May 6, 2020. YouTube. https://www.youtube.com/watch?v=unlQXGSNKOI

Chapter 16: "What would Gene Roberts have done?"

121. Mark Honigsbaum. *The Pandemic Century*. Chapter 4: "The Philly Killer." London: Hurst & Company, 2019.
122. Michael Capuzzo. The Drug that Cracked COVID. *Mountain Home*, May 2021. https://www.mountainhomemag.com/article_tags/michael-capuzzo
123. Michael Capuzzo. e-mail to John Leake, April 10, 2022.

Chapter 17: Nihilism and Fraud

124. Kary Mullis. 2013 Interview. https://www.bitchute.com/video/I34oouXAb7Dj/
125. M. Mehra et. al. Hydroxychloroquine or chloroquine with or without a macrolide for treatment of COVID-19: a multinational registry analysis. The *Lancet*, May 22, 2020. https://www.thelancet.com/journals/lancet/article/PIIS0140-6736(20)31180–6/fulltext
126. Catherine Offord. The Sugisphere Scandal: What Went Wrong? *The Scientist*, October 1, 2020. https://www.the-scientist.com/features/the-surgisphere-scandal-what-went-wrong—67955
127. Ibid.
128. Retraction—M. Mehra et. al. Hydroxychloroquine or chloroquine with or without a macrolide for treatment of COVID-19: a multinational registry analysis. the *Lancet*, June 5, 2020. https://www.thelancet.com/journals/lancet/article/PIIS0140-6736(20)31324–6/fulltext
129. Harvey Risch. "The Key to Defeating COVID-19 Already Exists. We Need to Start Using It | Opinion." *NEWSWEEK*, July 23, 2020. https://www.newsweek.com/key-defeating-COVID-19-already-exists-we-need-start-using-it-opinion-1519535
130. H. Risch. Early Outpatient Treatment of Symptomatic, High-Risk COVID-19 Patients that Should be Ramped-Up Immediately as Key to the Pandemic Crisis. *Am J Epidemiol.* 2020 May 27. https://www.ncbi.nlm.nih.gov/pmc/articles/PMC7546206/

Chapter 18: Professor Risch Punches Back

131. Dawn Connelly. A History of Aspirin. *The Pharmaceutical Journal*, Sept. 26, 2014. https://pharmaceutical-journal.com/article/infographics/a-history-of-aspirin
132. C. N. McAlister et al. The Halifax disaster (1917): eye injuries and their care. *Br J Ophthalmol.* 2007 Jun; 91(6): 832–835. https://www.ncbi.nlm.nih.gov/pmc/articles/PMC1955605
133. S. I. Hajdu et al. A Note from History: The Use of Tobacco. *Annals of Clinical & Laboratory Science*, vol. 40, no. 2, 2010 http://www.annclinlabsci.org/content/40/2/178.full.pdf
134. Knut-Olaf Haustein. Fritz Lickint (1898–1960) Ein Leben als Aufklaerer über die Gefahren des Tabaks. https://web.archive.org/web/20141105152951/http://www.ecomed-medizin.de/sj/sfp/Pdf/aId/6824
135. Larry Tye. *The Father of Spin: Edward L. Bernays and the Birth of Public Relations*. Henry Holt and Company, New York, 1998.
136. Ibid.
137. Edward Bernays. *Propaganda*. New York: Horace Liveright. 1928, pp. 9–10.
138. Reeves, B. et al. *Cochrane Training Manual.* Chapter 24: Including non-randomized studies on intervention effects. https://training.cochrane.org/handbook/current/chapter-24
139. S. Ebrahim et. al. Reanalyses of Randomized Clinical Trial Data. *JAMA.* 2014;312(10):1024–1032. https://jamanetwork.com/journals/jama/fullarticle/1902230
140. FDA, Real-World Evidence https://www.fda.gov/science-research/science-and-research-special-topics/real-world-evidence
141. RECOVERY. No clinical benefit from use of hydroxychloroquine in hospitalised patients with COVID-19. June 5, 2020. https://www.recoverytrial.net/news/statement-from-the-chief-investigators-of-the-randomised-evaluation-of-COVID-19-therapy

-recovery-trial-on-hydroxychloroquine-5-june-2020-no-clinical-benefit-from-use-of-hydroxychloroquine-in-hospitalised-patients-with-COVID-19

142. Meryl Nass, MD. COVID-19 Has Turned Public Health Into a Lethal, Patient-Killing Experimental Endeavor. *Alliance for Human Research Protection*, June 20, 2020. https://ahrp.org/COVID-19-has-turned-public-health-into-a-lethal-patient-killing-experimental-endeavor/

Chapter 20: Healers of the Imperial Valley

143. Tom Bodus. FORWARD THINKERS: Dr. George Fareed: Not your ordinary country doctor. *Valley Woman*, March 1, 2021. https://www.ivpressonline.com/valleywomen/forward-thinkers-dr-george-fareed-not-your-ordinary-country-doctor/article_29c47254-ff90-11eb-8c00-ab73a531f7f7.html

144. George Fareed, MD and Brian Tyson, MD. *Overcoming COVID Darkness: How Two Doctors Successfully Treated 7000 Patients*. Published by Brian Tyson, MD, and George Fareed, MD.

145. Brian Tyson, MD. The miracle of the Imperial Valley: Dr. Tyson's first-person account of COVID-19. *Desert Review*, Nov. 1, 2020. https://www.thedesertreview.com/news/the-miracle-of-the-imperial-valley-dr-tyson-s-first-person-account-of-COVID-19/article_a8707136-196b-11eb-bc7b-87d7730460bb.html

Chapter 21: "For the sake of our parents."

146. FDA. FDA cautions against use of hydroxychloroquine or chloroquine for COVID-19 outside of the hospital setting or a clinical trial due to risk of heart rhythm problems. July 1, 2020. https://www.fda.gov/drugs/drug-safety-and-availability/fda-cautions-against-use-hydroxychloroquine-or-chloroquine-COVID-19-outside-hospital-setting-or

147. Harvey Risch. The Key to Defeating COVID Already Exists. We Need to Start Using It. *NEWSWEEK*, July 23, 2020. https://www.newsweek.com/key-defeating-COVID-19-already-exists-we-need-start-using-it-opinion-1519535

148. Harvey Risch. The Author Replies. *American Journal of Epidemiology*, Volume 189, Issue 11, November 2020, Pages 1444–1449. https://academic.oup.com/aje/article/189/11/1444/5873640

149. Harvey Risch. The Key to Defeating COVID Already Exists. We Need to Start Using It. *NEWSWEEK*, July 23, 2020. https://www.newsweek.com/key-defeating-COVID-19-already-exists-we-need-start-using-it-opinion-1519535

Chapter 22: Enlightenment and Censorship

150. Peter A. McCullough, MD, et al. Pathophysiological Basis and Rationale for Early Outpatient Treatment of SARS-CoV-2 (COVID-19) Infection. *The American Journal of Medicine*, August 06, 2020DOI https://www.amjmed.com/article/S0002-9343(20)30673–2/fulltext

151. Treatment with Hydroxychloroquine Cut Death Rate Significantly in COVID-19 Patients. Henry Ford Health System Study Shows. *Henry Ford Health*, June 2, 2020. https://www.henryford.com/news/2020/07/hydro-treatment-study

152. NIH. Treatment Guidelines, August 27, 2020. https://files.COVID19treatmentguidelines.nih.gov/guidelines/archive/COVID19treatmentguidelines-08-27-2020.pdf
153. C. Chaccour, F. Hammann, S. Ramon-Garcia, N. R. Rabinovich. Ivermectin and COVID-19: keeping rigor in times of urgency. *Am J Trop Med Hyg*. 2020;102(6):1156–1157. https://www.ncbi.nlm.nih.gov/pubmed/32314704
154. C. A. Guzzo, C. I. Furtek, A. G. Porras, et al. Safety, tolerability, and pharmacokinetics of escalating high doses of ivermectin in healthy adult subjects. *J Clin Pharmacol.* 2002;42(10):1122–1133. https://www.ncbi.nlm.nih.gov/pubmed/12362927

Chapter 23: "We can beat COVID together."

155. NIH. Treatment Guidelines, October 9, 2020. https://files.COVID19treatmentguidelines.nih.gov/guidelines/archive/COVID19treatmentguidelines-10-09-2020.pdf
156. Katharine J. Wu and Gina Kolata. Remdesivir Fails to Prevent COVID-19 Deaths in Huge Trial. *New York Times,* Nov. 19, 2020. https://www.nytimes.com/2020/10/15/health/coronavirus-remdesivir-who.html
157. Francesco Gaurascio. World's top intensive care body advises against remdesivir for sickest COVID patients. *REUTERS,* Nov. 13, 2020. https://www.reuters.com/article/us-health-coronavirus-remdesivir-gilead-idINKBN27T13W
158. Ibid.
159. University of Oxford. Common asthma treatment reduces need for hospitalisation in COVID-19 patients, study suggests. Feb. 9, 2021. https://www.ox.ac.uk/news/2021-02-09-common-asthma-treatment-reduces-need-hospitalisation-COVID-19-patients-study

Chapter 25: Dr. McCullough Goes to Washington

160. US Senate Committee On Homeland Security & Government Affairs. Early Outpatient Treatment: An Essential Part of a COVID-19 Solution. Full Committee Hearing, Nov. 19, 2020. https://www.hsgac.senate.gov/hearings/early-outpatient-treatment-an-essential-part-of-a-COVID-19-solution

CHAPTER 26: The Empire Strikes Back

161. Ashish Jha, MD. The Snake Oil Salesmen of the Senate. *New York Times*, Nov. 24, 2020. https://www.nytimes.com/2020/11/24/opinion/hydroxychloroquine-COVID.html
162. Georgetown University Center for Global Health Science & Security Pandemic Preparedness in the Next Administration. January 10, 2017. https://ghss.georgetown.edu/pandemicprep2017/
163. Peter R. Breggin, MD, and Ginger Ross Breggin, *COVID-19 and the Global Predators: We Are the Prey*. Ithaca: Lake Edge Press, 2021, p. 259.

Chapter 27: Ivermectin Gets a Hearing

164. Katharine J. Wu No, the Pfizer and Moderna vaccine development has not been 'reckless.' *New York Times,* Dec. 8, 2020. https://www.nytimes.com/2020/12/08/technology/COVID-vaccines-senate-hearing.html

165. Cheryl Gay Stolberg. Anti-Vaccine Doctor Has Been Invited to Testify Before Senate Committee. *New York Times*, Dec. 6, 2020. https://www.nytimes.com/2020/12/06/us/politics/anti-vax-scientist-senate-hearing.html
166. US Senate Committee On Homeland Security & Government Affairs. Early Outpatient Treatment: An Essential Part of a COVID-19 Solution, Part II. Full Committee Hearing, Dec. 8, 2020. https://www.hsgac.senate.gov/early-outpatient-treatment-an-essential-part-of-a-COVID-19-solution-part-ii

Chapter 28: Begging for the Wonder Drug

167. Michael Capuzzo. The Drug that Cracked COVID. *Mountain Journal*, May 2021.
168. Author Interview with Beth Parlato. March 2, 2020.

Chapter 29: An Orgy of Federal Money

169. Nancy Ochieng et al. Funding for Health Care Providers During the Pandemic: An Update. Kaiser Family Foundation, Jan. 27, 2022. https://www.kff.org/coronavirus-COVID-19/issue-brief/funding-for-health-care-providers-during-the-pandemic-an-update/
170. Charles Creitz. Minnesota doctor blasts "ridiculous" CDC corona death count guidelines. *Fox News*, April 9, 2020. https://www.foxnews.com/media/physician-blasts-cdc-coronavirus-death-count-guidelines
171. CDC. Guidance for Certifying Deaths Due to Coronavirus Disease 2019 (COVID–19) Report No 3, April 2020. https://www.cdc.gov/nchs/data/nvss/vsrg/vsrg03-508.pdf
172. Zoe Chace. How Perverse Incentives Drive Up Health Care Costs. *NPR Morning Edition*, Jan. 16, 2014. https://www.npr.org/2014/01/16/262946913/how-perverse-incentives-drive-up-health-care-costs
173. Nancy Ochieng et al. Funding for Health Care Providers During the Pandemic: An Update. Kaiser Family Foundation. Jan. 27, 2022. https://www.kff.org/coronavirus-COVID-19/issue-brief/funding-for-health-care-providers-during-the-pandemic-an-update/
174. MacKenzie Bean. Gilead saw $5.6 in remdesivir sales last year. *Becker's Hospital Review*, Feb. 2, 2022. https://www.beckershospitalreview.com/pharmacy/gilead-saw-5-6b-in-remdesivir-sales-last-year.html
175. Joseph, Andrew. WHO group recommends against using remdesivir to treat hospitalized COVID-19 patients. *STAT*, Nov. 19, 2020. https://www.statnews.com/2020/11/19/who-recommends-against-remdesivir-COVID-19/

Chapter 30: "The best disguise is the truth."

176. Mike Magee, MD. *Code Blue: Inside America's Medical Industrial Complex*. New York: Grove Atlantic, 2019. Kindle Edition. (p. 19).
177. American Resistance to a Standing Army. Teaching History.org https://teachinghistory.org/history-content/ask-a-historian/24671
178. Carl Jung. "The Shadow." *Collected Works of C.G. Jung*. 9ii, par. 14.

179. Federal Criminal Practice Group of Price Benowitz LLP. Common Federal Conspiracy Cases in DC. https://whitecollarattorney.net/dc-federal-conspiracy-lawyer/common-cases/

Chapter 31: "I'm in a very sensitive position here."

180. Peter A. McCullough, MD, et al. Multifaceted highly targeted sequential multidrug treatment of early ambulatory high-risk SARS-CoV-2 infection (COVID-19). *Rev. Cardiovasc. Med.* 2020, 21(4), 517–530; Dec. 30, 2020. https://www.imrpress.com/journal/RCM/21/4/10.31083/j.rcm.2020.04.264/htm
181. A Letter to Andrew Hill, Dr. Tess Lawrie. Oracle Films. March 4, 2022. https://odysee.com/@OracleFilms:1/2022.03.04-A-Letter-to-Andrew-Hill-V8_HD:3
182. Ibid.
183. UNITAID. Unitaid funding sees launch of world's first long-acting medicines centre at University of Liverpool, Jan. 12, 2021. https://unitaid.org/news-blog/unitaid-funding-sees-launch-of-worlds-first-long-acting-medicines-centre-at-university-of-liverpool/#en
184. A Letter to Andrew Hill, Dr. Tess Lawrie. Oracle Films, March 4, 2022. https://odysee.com/@OracleFilms:1/2022.03.04-A-Letter-to-Andrew-Hill-V8_HD:3

Chapter 32: The Pied Piper of Science

185. The Portal, Ep. #25 (solo with host Eric Weinstein), The Construct—Jeffrey Epstein. https://www.youtube.com/watch?v=dJNjH4SP6vw
186. Emily Flitter and James B. Stewart. Bill Gates Met with Jeffrey Epstein Many Times, Despite His Past. *New York Times*, Oct. 12, 2019. https://www.nytimes.com/2019/10/12/business/jeffrey-epstein-bill-gates.html

Chapter 33: "Rest in peace, wheezy."

187. Timothy Bella. Jimmy Kimmel suggests hospitals shouldn't treat unvaccinated patients who prefer ivermectin. *Washington Post*, Sept. 8, 2021. https://www.washingtonpost.com/arts-entertainment/2021/09/08/jimmy-kimmel-hospitals-unvaccinated-ivermectin/
188. Justus R. Hope, MD. The great Ivermectin deworming hoax. *Desert Review*, Sept. 6, 2021. https://www.thedesertreview.com/opinion/columnists/the-great-ivermectin-deworming-hoax/article_19b8f2a6-0f29-11ec-94c1-4725bf4978c6.html
189. EVENT 201 Videos. https://www.centerforhealthsecurity.org/event201/videos.html
190. Ibid.
191. Mary Beth Pfeiffer. A Judge Stands up to a Hospital: "Step Aside" and Give a Dying Man Ivermectin. *RESCUE* with Michael Capuzzo, Nov. 23, 2021. https://rescue.substack.com/p/a-judge-stands-up-to-a-hospital-step?s=r
192. May Beth Pfeiffer. From Near Death in a Hospital to Happy at Home, a Father Is Saved by His Daughter and Ivermectin. *RESCUE* with Michael Capuzzo, Nov 29, 2021. https://rescue.substack.com/p/from-near-death-in-a-hospital-to?s=r
193. Jonathan Gardner. Merck sees up to $7B in coming sales of coronavirus pill. *BIOPHARMA DIVE*, October 28, 2021. https://www.biopharmadive.com/news/merck-coronavirus-pill-7-billion-sales/609064/

Chapter 34: "Where's the focus on sick people?"

194. Harris Richard. Tracking Down Antibody Treatment Is A Challenge For COVID-19 Patients. *NPR*, Jan. 11, 2021. https://www.npr.org/sections/health-shots/2021/01/11/955716308/tracking-down-antibody-treatment-is-a-challenge-for-COVID-19-patients
195. Peter A. McCullough, MD, testifies to Texas Senate HHS Committee. March 10, 2021. YouTube Channel, Association of American Physicians and Surgeons. https://www.youtube.com/watch?v=QAHi3lX3oGM

Chapter 35: *Tucker Carlson Today*

196. *Tucker Carlson Today*—Dr. Peter McCullough - May 7, 2021. YouTube. https://www.youtube.com/watch?v=F7cLxs8fNq8

Chapter 36: For the Love of Money

197. Mike Magee, MD. *Code Blue: Inside America's Medical Industrial Complex*. New York: Grove Atlantic, 2019.
198. Joe Stephens. Where Profits and Lives Hang in Balance. *Washington Post*, Dec. 17, 2000. https://www.washingtonpost.com/archive/politics/2000/12/17/where-profits-and-lives-hang-in-balance/90b0c003-99ed-4fed-bb22-4944c1a98443/
199. John Le Carré. *The Constant Gardener*. New York: Scribner, Kindle Edition, p. 548.
200. Gardiner Harris. Pfizer Pays $2.3 Billion to Settle Marketing Case. *New York Times*, Sept. 2, 2009. https://www.nytimes.com/2009/09/03/business/03health.html
201. US Department of Justice. Justice Department Announces Largest Health Care Fraud Settlement in Its History: Pfizer to Pay $2.3 Billion for Fraudulent Marketing. Sept. 2, 2009. https://www.justice.gov/opa/pr/justice-department-announces-largest-health-care-fraud-settlement-its-history
202. Mike Magee, MD. *Code Blue: Inside America's Medical Industrial Complex*. New York: Grove Atlantic, 2019, Chapter 8 "Masters of Manipulation."

Chapter 37: The Conscience of Psychiatry

203. Jane B. Baird. Mindbending Controversy. *The Harvard Crimson*, January 16, 1974. https://www.thecrimson.com/article/1974/1/16/mindbending-controversy-pimost-neurosurgeons-now-agree/

Chapter 38: Empire of Pain

204. Patrick Radden Keefe. The Family that Built and Empire of Pain. *New Yorker*, Oct. 30, 2017. https://www.newyorker.com/magazine/2017/10/30/the-family-that-built-an-empire-of-pain
205. Mike Magee, MD. *Code Blue: Inside America's Medical Industrial Complex*. New York: Academic Monthly Press, 2019, Chapter 8, "Masters of Manipulation."

206. Patrick Radden Keefe. The Family that Built and Empire of Pain. *New Yorker*, Oct. 30, 2017. https://www.newyorker.com/magazine/2017/10/30/the-family-that-built-an-empire-of-pain
207. CDC, Drug Overdose Deaths in the U.S. Top 100,000 Annually. https://www.cdc.gov/nchs/pressroom/nchs_press_releases/2021/20211117.htm

Chapter 39: The Philosopher

208. John Leake. Dr. Peter McCullough Interview. VIMEO. May 19, 2021. https://vimeo.com/553518199

Chapter 41: The Stripping

209. Baylor health sues COVID-19 Vaccine skeptic, demands he stop using its name. *Dallas Morning News*, July 29, 2021. https://www.dallasnews.com/news/courts/2021/08/13/baylor-health-sues-COVID-19-vaccine-skeptic-and-demands-dallas-doctor-stop-using-its-name/

Chapter 42: Other Assassinations

210. Senator Ron Johnson. YouTube Channel. COVID-19: A Second Opinion, January 24, 2022. https://www.youtube.com/watch?app=desktop&v=9jMONZMuS2U

Chapter 43: *The Joe Rogan Experience*

211. The Joe Rogan Experience. Episode #1718—Dr. Sanjay Gupta. Spotify, Oct. 2021 https://open.spotify.com/episode/6rAgS1KiUvLRNP4HfUePpA
212. The Joe Rogan Experience. Episode #1747—Dr. Peter A. McCullough, MD. Spotify, Dec. 2021. https://open.spotify.com/episode/0aZte37vtFTkYT7b0b04Qz
213. The Joe Rogan Experience. Episode. Episode #1757—Dr. Robert Malone, MD. Spotify, Dec. 2021. https://open.spotify.com/episode/3SCsueX2bZdbEzRtKOCEyT
214. John Milton. "Areopagitica; A speech of Mr. John Milton for the Liberty of Unlicenc'd Printing, to the Parlament of England."

Chapter 44: Keeping Our Noses Clean

215. Md. Iqbal Mahmud Choudhury et al. Effect of 1% Povidone Iodine Mouthwash/Gargle, Nasal and Eye Drop in COVID-19 patient. *Bioresearch Communications-(BRC)*. Dhaka, Bangladesh, 7(1), pp. 919–923. Available at: https://www.bioresearchcommunications.com/index.php/brc/article/view/176
216. L. Zou et al., (2020). SARS-CoV-2 Viral Load in Upper Respiratory Specimens of Infected Patients. The New England Journal of Medicine, 382(12), 1177–1179. https://www.nejm.org/doi/10.1056/NEJMc2001737

217. N. F. Farrell et al. (2020). Benefits and Safety of Nasal Saline Irrigations in a Pandemic—Washing COVID-19 Away. *JAMA* Otolaryngol Head Neck Surg., 146(9):787–788. https://jamanetwork.com/journals/jamaotolaryngology/fullarticle/2768627
218. T. Kontiokari, M. Uhari, M. Koskela. (1998). Antiadhesive effects of xylitol on otopathogenic bacteria. The Journal of Antimicrobial Chemotherapy, 41(5), 563–565. https://academic.oup.com/jac/article/41/5/563/652694?login=false
219. G. Ferrer et al. (2020). A Nasal Spray Solution of Grapefruit Seed Extract plus Xylitol Displays Virucidal Activity Against SARS-Cov-2 In Vitro. BioRxiv. https://www.biorxiv.org/content/10.1101/2020.11.23.394114v1
220. J. C. Vega et al. (2020). Iota carrageenan and xylitol inhibit SARS-CoV-2 in Vero cell culture. BioRxiv. https://www.biorxiv.org/content/10.1101/2020.08.19.225854v1
221. Ashley Abramson. Why Doctors Warn Against Using Betadine to Prevent COVID-19, Health, May 22, 2022. https://www.health.com/condition/infectious-diseases/coronavirus/betadine-nasal-spray-mouthwash-COVID

Chapter 45: "The best investment I've ever made."

222. Medscape. Worst of 2021. https://www.medscape.com/slideshow/2021-physicians-of-the-year-6014684#26
223. Bill Gates. TED 2010. https://www.youtube.com/watch?v=JaF-fq2Zn7I
224. Martin Enserink. Gates Calls for "Decade of Vaccines," Pledges Assault on Child Mortality. *Nature*, January 29, 2010. https://www.science.org/content/article/gates-call-decade-vaccines-pledges-assault-child-mortality
225. Bill & Melinda Gates Foundation. Global Health Leaders Launch Decade of Vaccines Collaboration | Bill & Melinda Gates Foundation. https://www.gatesfoundation.org/ideas/media-center/press-releases/2010/12/global-health-leaders-launch-decade-of-vaccines-collaboration
226. Matthew J. Belvedere. Bill Gates: My "best investment" turned $10 billion into $200 billion worth of economic benefit. *CNBC*, January 23, 2019. https://www.cnbc.com/2019/01/23/bill-gates-turns-10-billion-into-200-billion-worth-of-economic-benefit.html
227. Bill Gates. The Best Investment I've Ever Made. *Wall Street Journal*, Jan. 16, 2010. https://www.wsj.com/articles/bill-gates-the-best-investment-ive-ever-made-11547683309
228. Best of Bill Gates Being Deposed by David Boies United States v. Microsoft Corp.1998–2001. YouTube. https://www.youtube.com/watch?v=gRelVFm7iJE
229. John Naughton. Let's not forget, Bill Gates hasn't always been the good guy. *Guardian*, August 29, 2020. https://www.theguardian.com/commentisfree/2020/aug/29/lets-not-forget-bill-gates-hasnt-always-been-the-good-guy
230. Paul Thurrott. Judge Jackson Exits Microsoft Discrimination Case. *ITPro Today*, March 13, 2001. https://www.itprotoday.com/windows-78/judge-jackson-exits-microsoft-discrimination-case
231. Richard Jensen. Bill Gates—Another Rockefeller or Another Ford? *OSU-EDU ORIGINS*, June 2000. https://origins.osu.edu/history-news/bill-gates-another-rockefeller-or-another-ford?language_content_entity=en
232. John Vidal. Are Gates and Rockefeller using their influence to set agenda in poor states? *Guardian*, January 15, 2016. https://www.theguardian.com/global-development/2016/jan/15/bill-gates-rockefeller-influence-agenda-poor-nations-big-pharma-gm-hunger

233. Michael Barbaro. Can Bill Gates Vaccinate the World? How the Microsoft founder is changing the way the world is vaccinated and, potentially the course of the pandemic. *New York Times* podcast (Mar. 3, 2021). https://www.nytimes.com/2021/03/03/podcasts/the-daily/coroanvirus-vaccine-bill-gates-covax.html?showTranscript=1
234. Rebecca G. Baker. Bill Gates Asks NIH Scientists for Help in Saving Lives And Explains Why the Future Depends on Biomedical Innovation. *THE NIH CATALYST* (Jan-Feb, 2014). https://irp.nih.gov/catalyst/v22i1/bill-gates-asks-nih-scientists-for-help-in-saving-lives
235. Tim Schwab. Journalism's Gates Keepers. *Columbia Journalism Review*, Aug. 21, 2020. https://www.cjr.org/criticism/gates-foundation-journalism-funding.php
236. Erin Fox. How Pharma Companies Game the System to Keep Drugs Expensive. *Harvard Business Review*, April 6, 2017. https://hbr.org/2017/04/how-pharma-companies-game-the-system-to-keep-drugs-expensive
237. G. H. Jones et al. Strategies that delay or prevent the timely availability of affordable generic drugs. *Blood*, 2016 Mar 17; 127(11): 1398–1402.
238. Julie Margetta Morgan and Devin Duffy. The Cost of Capture: How the Pharmaceutical Industry has Corrupted Policy and Harmed Patients. *The Roosevelt Institute*, May 2019. https://rooseveltinstitute.org/wp-content/uploads/2020/07/RI_Pharma_Cost-of-Capture_brief_201905.pdf
239. Sheila Kaplan et al. Trump's Vaccine Chief Has Vast Ties to Drug Industry, Posing Possible Conflicts. *New York Times*, May 20, 2020. https://www.nytimes.com/2020/05/20/health/coronavirus-vaccine-czar.html
240. WEF. The Great Reset. https://www.weforum.org/focus/the-great-reset

Chapter 46: Bringing America Home

241. Jack Bingham. Bayer executive: mRNA shots are "gene therapy" marketed as "vaccines" to gain public trust. *LIFESITE*, Nov. 10, 2021. https://www.lifesitenews.com/news/bayer-executive-mrna-shots-are-gene-therapy-marketed-as-vaccines-to-gain-public-trust/
242. Washington's Farewell Address. https://www.govinfo.gov/content/pkg/GPO-CDOC-106sdoc21/pdf/GPO-CDOC-106sdoc21.pdf
243. Friedrich Schiller. "The Song of the Bell."

Selected Bibliography

The following is a list of works by Dr. Peter McCullough and his colleagues on the early treatment of COVID-19. The papers present evidence of the efficacy of various drugs and supplements for preventing and treating the disease—evidence that Dr. McCullough and his colleagues were able to ascertain and assemble as the events documented in this book were unfolding. When SARS-CoV-2 arrived, they set about trying to discover how to fight it. Their papers are a record of how early treatments were discovered and improved over time. The essence of their discovery is that COVID-19 is a *treatable* disease, especially when it is tackled early.

Papers by Peter A. McCullough, MD, MPH

Peter A. McCullough, MD, et al. "Pathophysiological Basis and Rationale for Early Outpatient Treatment of SARS-CoV-2 (COVID-19) Infection." *The American Journal of Medicine*, August 6, 2020.

Peter A. McCullough, MD, et al. "Multifaceted highly targeted sequential multidrug treatment of early ambulatory high-risk SARS-CoV-2 infection (COVID-19)." *Rev. Cardiovasc. Med.* 2020, 21(4), 517–530; December 30, 2020.

Peter A. McCullough, Raphael B. Stricker, Harvey A. Risch. "Role of hydroxychloroquine in multidrug treatment of COVID-19." *Reviews in Cardiovascular Medicine*, Volume 22, Issue 3, September 24, 2021.

Brian C. Proctor, Peter A. McCullough, et al. "Clinical outcomes after early ambulatory multidrug therapy for high-risk SARS-CoV-2 (COVID-19) infection." *Reviews in Cardiovascular Medicine*, Volume 21, Issue 4, 611–614. November 25, 2020.

Papers by Other Authors Mentioned in this Book

Matthieu Million, Didier Raoult, et al. "Early combination therapy with hydroxychloroquine and azithromycin reduces mortality in 10,429 COVID-19 outpatients." *Reviews in Cardiovascular Medicine*, Volume 22, Issue 3, 1063–1072; September 24, 2021.

Pierre Kory, Paul Marik, et al. "Review of the Emerging Evidence Demonstrating the Efficacy of Ivermectin in the Prophylaxis and Treatment of COVID-19." *American Journal of Therapeutics*, Volume 28 - Issue 3 - p e299-e318. May/June 2021.

FLCCC ALLIANCE. One Page Summary of the Clinical Trials Evidence for Ivermectin in COVID-19 as of January 11, 2021.

JC Rajter et al. "Use of Ivermectin Is Associated With Lower Mortality in Hospitalized Patients With Coronavirus Disease 2019." *Chest Infections: Original Research*, Volume 159, Issue 1, P85-92, January 1, 2021.

Harvey Risch. "Early Outpatient Treatment of Symptomatic, High-Risk COVID-19 Patients that Should be Ramped-Up Immediately as Key to the Pandemic Crisis." *Am J Epidemiol.* May 27, 2020.

Harvey Risch. "The Author Replies." *Am J Epidemiol.*, Volume 189, Issue 11, November 2020, Pages 1444–1449.

Risch, Harvey. "The Key to Defeating COVID Already Exists. We Need to Start Using It." *Newsweek*, July 23, 2020.

Roland Derwand, Martin Scholz, Vladimir Zelenko. "COVID-19 outpatients: early risk-stratified treatment with zinc plus low-dose hydroxychloroquine and azithromycin: a retrospective case series study." *International Journal of Antimicrobial Agents.* Volume 56, Issue 6, December 2020.

Paper Vindicating Dr. Richard Bartlett's Case Study on Efficacy of Inhaled Budesonide

Sanjay Ramakrishnan, MBBS, et al. "Inhaled budesonide in the treatment of early COVID-19 (STOIC): a phase 2, open-label, randomized controlled trial." *Lancet*, Volume 9, Issue 7, Pages 763–772. July 1, 2021.

Index

Acknowledgments

From Peter:
I am immeasurably grateful for the unwavering support of my wife, Maha, whose world has been rocked so many times since I embarked on this journey. She has been the steadiest and most stalwart of companions, and I am so fortunate to be married to her. Our children, Haley and Sean, have watched the dismantling of my career in real time and never questioned my convictions or reasoning. I am grateful for the understanding and grace of my parents, Mary and Thomas. Maha's parents, Karima and Hilmi, have also shown great support in the midst of their own innumerable trials at their advanced age.

My heartfelt thanks to the thousands of physicians, nurses, health professionals, and scientists who have reached out to me to assist and collaborate in saving as many lives as possible all over the world. There is not enough space in this or any other tome to express my gratitude for them. Prior to the pandemic, I devoted my entire life to studying and practicing medicine. I never imagined that I would one day become a public figure, constantly in the media. Within the US legacy media, Tucker Carlson and Laura Ingraham at Fox News have demonstrated an exceptional open-mindedness and spirit of inquiry in their conversations with me. Joe Rogan was as intrepid as ever, first by inviting me to join him in conversation, and then by sticking with his free speech guns in the face of considerable pressure to censor me.

Within the new, independent media space, I am grateful for the warm welcome I received from innumerable coordinators, producers, and reporters who have shown keen interest in my work. If it weren't for them, I would not have been able to communicate with the world. Special thanks to Malcolm, America Out Loud; Daniel O'Connor, TrialSiteNews; Daniel

Horowitz, Blaze Media; Joni and the late Marcus Lamb, Daystar Television Network; Alex Jones, InfoWars; Chris Salcedo, NewsMax; Eric Bolling, The Balance, NewsMax; Gene Bailey, FlashPoint, Victory Channel; Jan Jekielak, The Epoch Times; Dr. Sebastian Gorka, America First; Brannon Howse, Lindell-TV; Sean Spicer, Spicer & Co.; NewsMax; Mat Staver, Liberty Counsel Network; and the young podcaster, Tommy Carrigan, the TCP Podcast.

Finally, I would like to give a very special thanks to Senator Ron Johnson (R-WI). In my view, he is the finest US senator in history for his courageous stand on behalf of suffering patients and compassionate doctors who fought against all odds during this crisis.

From John:
I would like to give special thanks to treating physicians Drs. Vladimir Zelenko and Ivette Lozano for taking time out of their busy schedules to tell me their fascinating stories. Dr. Zelenko was kind enough to visit me in Dallas for two days. Professor Harvey Risch spent the better part of a Saturday with me, explaining the "aspects of causal reasoning" and just about everything else in the world. Yale is immeasurably lucky to have him on its staff.

Among patients, Adan and Roxane Gonzalez spent a whole day telling me their gripping story. Jodi Carroll took great pains to tell me what it's like and how it feels to fight for a dying patient to whom treatment is inexplicably denied. Attorney Beth Parlato offered a vivid account of what it's like to face the Bio-Pharmaceutical Complex in court. I would also like to acknowledge the work of my fellow investigative authors who have researched this labyrinthine story and happily shared their findings with me. Robert F. Kennedy Jr.'s book *The Real Anthony Fauci* has been an invaluable reference work, especially its chapter on remdesivir. Likewise, Dr. Peter Breggin did extraordinary work in researching the "Global Predators" who are the main perpetrators in this true crime story. Dr. McCullough shared his insights with both authors, and they have shared their insights with us. Michael Capuzzo and Mary Beth Pfeiffer have been very generous in sharing the fruits of their research.

Finally, I would like to give special thanks to my mother, Kathy Leake, who has been a lifelong and tireless supporter of my education and work.

About the Authors

John Leake studied history and philosophy with Roger Scruton at Boston University. He then went to Vienna, Austria, on a graduate school scholarship and ended up living in the city for over a decade, working as a freelance writer and translator. He is a true crime writer with a lifelong interest in medical history and forensic medicine.

Peter A. McCullough, MD, MPH, is an internist, cardiologist, and epidemiologist who has been a leader in the medical response to COVID-19. In August of 2020 he published "Pathophysiological Basis and Rationale for Early Outpatient Treatment of SARS-CoV-2 (COVID-19) Infection" in the *American Journal of Medicine*. This paper presented the first synthesis of sequenced multidrug treatment of ambulatory patients infected with the virus and was subsequently updated in *Reviews in Cardiovascular Medicine*. He has published dozens of peer-reviewed papers on the infection and has commented extensively on the medical response in The Hill, America Out Loud, ABC, Fox News, OAN, Newsmax, Victory Channel, and a multitude of independent media outlets. He has testified on several occasions in the US Senate and before several State Senate and Congressional Committees.